"[*The Myth of Human Supremacy*] offers a ne\\ of humans in relation to all other life on Ea most basic assumptions about human domin_____ ___ ___ _____ George Wuerthner, author, ecologist, and wildlands advocate

"This book dissects and demolishes one of our culture's most pernicious assumptions, that humans are the pinnacle of evolution and the supreme species on the planet. Derrick Jensen is a master at digging into our beliefs, turning over rocks and unflinchingly looking at what lies beneath. *The Myth of Human Supremacy* brilliantly exposes our dangerous, nature-devouring belief that humans are superior and reveals to what absurd lengths we will go to preserve that belief. This is an important book full of critical lessons. It shows the value—and urgency—of humbly taking our true, unexceptional but valuable place among all of life's marvelous creatures." —Toby Hemenway, author of *Gaia's Garden* and *The Permaculture City*

"When I read *Endgame* (2006), I believed I had found the clearest description of patriarchal civilization and how it is killing every aspect of the living planet. I was mistaken. Derrick Jensen has outdone himself. In heartfelt, compelling prose, he asks the reader to question the obvious lies embedded within the dominant paradigm." —Guy McPherson, professor emeritus of conservation biology at the University of Arizona

"Jensen's arguments are ferocious, heartbroken, hilarious, and lethally logical. The truths he tells are the most important in this reeling world, bar none." —Kathleen Dean Moore, author of *Moral Ground* and *Great Tide Rising*

"This book made me weep. It's an angry ballad, an anguished love song to life itself. I sit here, tears in my eyes as I type these words, as if yet another human needed to be heard from. I sit here wishing, dreaming we could instead hear what the Amani flatwing damselflies, ploughshare tortoises, Asiatic black bears, and the pea plants have to say about The Myth of Human Supremacy. I imagine they'd bellow in unison: 'It's about fuckin' time you caught on!'" —Mickey Z., author of *Occupy These Photos*

"Brilliant, lucid and gorgeously written, *The Myth of Human Supremacy* attacks the core of the planet-scale problem, the idea that only humans matter. The book is elegant and poised; the argument unassailable; the narrative engaging, witty, and full of surprises; the research meticulous. This is perhaps my favorite of his books." —Suprabha Seshan, environmental educator, activist, and restoration ecologist, winner of 2006 Whitley Fund for Nature award, Ashoka Fellow, executive director of Gurukula Botanical Sanctuary

"In this important book, Jensen upends longstanding 'truths' about human domination of the planet, demanding that we not only rethink our ideas about politics and economics, but about ourselves. He focuses our attention on the multiple, cascading crises that can be traced to human supremacy—the deeply destructive illusion that the world was made for humans because we are so very special. Jensen considers, and rejects, every reason we want to believe ourselves the anointed species, and challenges all of us to take seriously the moral principles we claim to hold." —Robert Jensen, University of Texas at Austin, author of *Plain Radical*

"*The Myth of Human Supremacy* is poetic and deeply moving. Jensen is unafraid to interrogate unquestionable assumptions and ask 'crazy' questions. Here he dismantles the core of our crises, the mythologies that guide authoritarian, unsustainable, human supremacist cultures. Read this and weep, but then with new awareness shake off emotional and ideological blinders you have been taught, and take action with those who understand that humans are one among many." —Darcia Narvaez, professor of psychology at the University of Notre Dame, blogger at *Psychology Today* ("Moral Landscapes"), and author of *Neurobiology and the Development of Human Morality: Evolution, Culture and Wisdom*

"Derrick Jensen elegantly shows that everything in our world is interconnected, and animals, plants, and even bacteria are sentient, conscious, and much like us. We humans refuse to believe that, preferring to believe a vast gulf exists between us and the rest of the natural world. That leads to the end of us and all of nature as we kill our planet. I hope this book will help people change their belief in human supremacy and help save our world." —Con Slobodchikoff, PhD, author of *Chasing Doctor Dolittle: Learning the Language of Animals*

"In his most important work since *A Language Older Than Words*, Jensen lays bare the sociopathy of the ideology of human supremacy: the fact that western 'civilization' is based on domination, thievery, and murder, while the natural world innately gravitates towards harmony and balance. This supremacy is destroying the planet, an infinitely complex living entity we've only barely begun to understand. This book is mandatory reading." —Dahr Jamail, author/journalist

"It is said that a revolution begins in the mind—an alternative to our present circumstances must first be imagined before we can be moved to fight for it. So we should all be grateful to Derrick Jensen, who with this book breaks the ideological chains of human supremacy and reveals the world as the interconnected web of being that it truly is. With our illusions ripped away, we may yet be able to save ourselves and our beautiful planet from the system that is killing us all." —Stephanie McMillan, author of *Capitalism Must Die*

THE MYTH OF HUMAN SUPREMACY

DERRICK JENSEN

SEVEN STORIES PRESS

New York • Oakland

Portions of this book appeared in different form in *Orion* and *Earth Island Journal.*

A SEVEN STORIES PRESS FIRST EDITION

Seven Stories Press
140 Watts Street
New York, NY 10013
sevenstories.com

Library of Congress Cataloging-in-Publication Data

Names: Jensen, Derrick, 1960- author.
Title: The myth of human supremacy / by Derrick Jensen.
Description: First edition. | New York, NY : Seven Stories Press, 2016. |
    Includes bibliographical references.
Identifiers: LCCN 2015046725| ISBN 9781609806781 (paperback) | ISBN
    9781609806798 (ebook)
Subjects: LCSH: Environmentalism. | Environmental ethics. | Nature--Effect of
    human beings on. | Human geography. | BISAC: NATURE / Environmental
    Conservation & Protection. | NATURE / Ecology.
Classification: LCC GE195 .J47 2016 | DDC 304.2--dc23
LC record available at http://lccn.loc.gov/2015046725

Printed in the United States of America

9 8 7 6 5 4 3 2 1

Usually, the English language reserves the pronoun *who* for humans and uses *that* for nonhumans. To align grammar and syntax with the ideas put forward in this book, many entities normally considered things will be referred to with the pronoun *who*.

—DJ

CONTENTS

For the earth

Just because some of us can read and write and do a little math, that doesn't mean we deserve to conquer the Universe.

—KURT VONNEGUT

PRELUDE

How we behave in the world is profoundly influenced by how we experience the world, which is profoundly influenced by how we perceive the world, which is profoundly influenced by what we believe about the world.

Our collective behavior is killing the planet.

It's not altogether irrelevant, then, to ask what sorts of beliefs (perceptions, experiences) might be leading to these destructive behaviors, and to ask how we can change these beliefs such that we will stop, not further, the murder of the planet.

We have been taught, in ways large and small, religious and secular, that life is based on hierarchies, and that those higher on these hierarchies dominate those lower, either by right or by might. We have been taught that there are myriad literal and metaphorical food chains where the one at the top is the king of the jungle.

But what if the point is not to rule, but to participate? What if life less resembles the board games *Risk* or *Monopoly*, and more resembles a symphony? What if the point is not for the violin players to drown out the oboe players (or worse, literally drown them or at least drive them from the orchestra, and take their seats for more violin players to use), but to make music with them? What if the point is for us to attempt to learn our proper role in this symphony, and then play that role?

—DJ, Northern California, January 2016

HUMAN SUPREMACISM

The modern conservative [and, I would say, the human suprema-cist] is engaged in one of man's oldest exercises in moral philosophy; that is, the search for a superior moral justification for selfishness.
JOHN KENNETH GALBRAITH

I'm sitting by a pond, in sunlight that has the slant and color of early fall. Wind blows through the tops of second-growth redwood, cedar, fir, alder, willow. Breezes make their way down to sedges, rushes, grasses, who nod their heads this way and that. Spider silk glistens. A dragonfly floats a few inches above the water, then suddenly climbs to perch atop a rush.

A family of jays talks among themselves.

I smell the unmistakable, slightly sharp scent of redwood duff, and then smell also the equally unmistakable and also slightly sharp, though entirely different, smell of my own animal body.

A small songbird, I don't know who, hops on two legs just above the waterline. She stops, cocks her head, then pecks at the ground.

Movement catches my eye, and I see a twig of redwood needles fall gently to the ground. It helped the tree. Now it will help the soil.

Someday I am going to die. Someday so are you. Someday both you and I will feed—even more than we do now, through our sloughed skin, through our excretions, through other means—those communities who now feed us. And right now, amidst all this beauty, all this life, all these others—sedge, willow, dragonfly, redwood, spider, soil, water, sky, wind, clouds—it seems not only ungenerous, but ungrateful to begrudge the present and future gift of my own life to these others without whom neither I nor this place would be who we are, without whom neither I nor this place would even be.

❀ ❀ ❀

Likewise, in this most beautiful place on Earth—and you do know, don't you, that each wild and living place on Earth is the most beautiful place on Earth—I can never understand how members of the dominant culture could destroy life on this planet. I can never understand how they could destroy even one place.

❀ ❀ ❀

Last year someone from *Nature* [sic] online journal interviewed me by phone. I include the *sic* because the journal has far more to do with promoting human supremacism—the belief that humans are separate from and superior to everyone else on the planet—than it has to do with the real world. Here is one of the interviewer's "questions": "Surely nature can only be appreciated by humans. If nature were to cease to exist, nature itself would not notice, as it is not conscious (at least in the case of most animals and plants, with the possible exception of the great apes and cetaceans) and, other than through life's drive for homeostasis, is indifferent to its own existence. Nature thus only achieves worth through our consciously valuing it."

At the precise moment he said this to me, I was watching through my window a mother bear lying on her back in the tall grass, her two children playing on her belly, the three of them clearly enjoying each other and the grass and the sunshine. I responded, "How dare you say these others do not appreciate life!" He insisted they don't.

I asked him if he knew any bears personally.

He thought the question absurd.

This is why the world is being murdered.

❀ ❀ ❀

Unquestioned beliefs are the real authorities of any culture.[1] A central unquestioned belief of this culture is that humans are superior to and separate from everyone else. Human supremacism is part of the foundation of much of this culture's religion,[2] science, economics, philosophy, art, epistemology, and so on.

Human supremacism is killing the planet. Human supremacists—at this point, almost everyone in this culture—have shown time and again that the maintenance of their belief in their own superiority, and the entitlement that springs from this belief, are more important to them than the well-being or existences of everyone else. Indeed, they've shown that the maintenance of this self-perception and entitlement are more important than the continuation of life on the planet.

Until this supremacism is questioned and dismantled, the self-perceived entitlement that flows from this supremacism guarantees that every attempt to stop this culture from killing the planet will fail, in great measure because these attempts will be informed and limited by this supremacism, and thus will at best be ways to slightly mitigate harm, with the primary point being to make certain to never in any way question or otherwise endanger the supremacism or entitlement.

In short, people protect what's important to them, and human supremacists have shown time and again that their sense of superiority and the tangible benefits they receive because of their refusal to perceive others as anything other than inferiors or resources to be exploited is more important to them than not destroying the capacity of this planet to support life, including, ironically, their own.

※ ※ ※

Especially because human supremacism is killing the planet, but also on its own terms, human supremacism is morally indefensible. It is also intellectually indefensible. Neither of which seems to stop a lot of people from trying to defend it.

The first line of defense of human supremacism is no defense at all, literally. This is true for most forms of supremacism, as unquestioned assumptions form the most common base for any form of bigotry: *Of course humans* (men, whites, the civilized) *are superior, why do you ask?* Or more precisely: *How could you possibly ask?* Or even more precisely: *What the hell are you talking about, you crazy person?* Or more precisely yet, an awkward silence while everyone politely forgets you said anything at all.

Think about it: if you were on a bus or in a shopping mall or in a church or in the halls of Congress, and you asked the people around you if they think humans are more intelligent than or are otherwise superior to cows or willows or rivers or mushrooms or stones ("stupid as a box of rocks"), what do you think people would answer? If you said to them that trees told you they don't want to be cut down and made into 2x4s, what would happen to your credibility? Contrast that with the credibility given to those who state publicly that you can have infinite economic (or human population) growth on a finite planet, or who argue that the world consists of resources to be exploited. If you said to people in this culture that oceans don't want to be murdered, would these humans listen? If you said that prairie dogs are in no way inferior to (or less intelligent than) humans, and you said this specifically to those humans who have passed laws *requiring* landowners to kill prairie dogs, would they be more likely to laugh at you or agree with you? Or do you think they'd be more likely to get mad at you? And just think how mad they'd get if you told them that land doesn't want to be owned (most especially by them). If you told them there was a choice between electricity from dams and the continued existence of salmon, lampreys, sturgeon, and mussels, which would they choose? Why? What are they already choosing?

❀ ❀ ❀

This is too abstract. Here is human supremacism. Right now in Africa, humans are placing cyanide wastes from gold mines on salt licks and in ponds. This cyanide poisons all who come there, from elephants to lions to hyenas to the vultures who eat the dead. The humans do this in part to dump the mine wastes, but mainly so they can sell the ivory from the murdered elephants.

Right now a human is wrapping endangered ploughshares tortoises in cellophane and cramming them into roller bags to try to smuggle them out of Madagascar and into Asia for the pet trade. There are fewer than 400 of these tortoises left in the wild.

Right now in China, humans keep bears in tiny cages, iron vests around the bears' abdomens to facilitate the extraction of bile from the bears' gall bladders. The bears are painfully "milked" daily. The vests also serve to keep the bears from killing themselves by punching themselves in the chest.

Right now there are fewer than 500 Amani flatwing damselflies left in the world. They live along one stream in Tanzania. The rest of their home has been destroyed by human agriculture.

This year has seen a complete collapse of monarch butterfly populations in the United States and Canada. Their homes have been destroyed by agriculture.

Right now humans are plowing under and poisoning prairies. Right now humans are clearcutting forests. Right now humans are erecting mega-dams. Right now because of dams, 25 percent of all rivers no longer reach the ocean.

And most humans couldn't care less.

Right now the University of Michigan Wolverines football team is hosting the Minnesota Golden Gophers. More than 100,000 humans are attending this football game. More than 100,000 humans have attended every Michigan home football game since 1975. There used to be real wolverines in Michigan. One was sighted there in 2004, the first time in 200 years. That wolverine died in 2010.

More people in Michigan—"The Wolverine State"—care about the Michigan Wolverines football team than care about real wolverines.

This is human supremacism.

※ ※ ※

I just got a note from a friend who was visiting her son. She writes, "Yesterday morning when I emptied the compost bucket, the guy next door called out to ask if that was 'garbage' I was putting on the pile. I told him it was 'compost.' We went back and forth a couple of times. Then he said, 'We don't want no [sic] animals around here. I saw a raccoon out there. There were never any animals around here before.' What better statement of human supremacism?"

※ ※ ※

Recently, scientists discovered that some species of mice love to sing. They "fill the air with trills so high-pitched that most humans can't even hear them." If "the melody is sweet enough, at least to the ears of a female mouse, the vocalist soon finds himself with a companion."

Mice, like songbirds, have to be taught how to sing. This is culture, passed from generation to generation. If they aren't taught, they can't sing.

So, what is the response by scientists to these mice, who love to sing, who teach each other how to sing, who sing for their lovers, who have been compared to "opera singers"?

Given what the ideology of human supremacism does to people who otherwise seem sane, we shouldn't be surprised to learn that the scientists wanted to find out what would happen if they surgically deafened these mice. And we shouldn't be surprised to learn that the mice could no longer sing their operas, their love songs. The deafened mice could no longer sing at all. Instead, they screamed.[3]

And who could blame them?

This is human supremacism.

※ ※ ※

Or there's this. Just yesterday I spoke with Con Slobodchikoff, who has been studying prairie dog language for more than thirty years. Through observing prairie dogs non-intrusively in the field, he has learned of the complexity of their language and social lives. But he has done so, he said, without the aid of grants. Time and again he was told that if he wanted to receive money for his research—and if he wanted to do "real science" instead of "just" observing nature—he would have to capture some prairie dogs, deafen them, and then see how these social creatures with their complex auditory language and communal relationships responded to their loss of hearing. Of course he refused. Of course he didn't receive the grants.

This is human supremacism.

※ ※ ※

And then today I got an email from a botanist friend who has worked for various federal agencies. His work has included identifying previously unknown species of plants. He said this work has not been supported by the agencies, because the existence of rare plants would interfere with their management plans, including the mass spraying of herbicides. His discoveries have been made on his own time and on his own dime.

It's a good thing science is value free, isn't it?

I told him Slobodchikoff had said to me that the scientific establishment makes it very difficult for people to manifest their love of the world. Slobodchikoff said this as someone who loves the earth very much.

My botanist friend agreed. "Science makes it very hard to love the world. Most scientists want the world to fit nice, clear, linear equations, and anything that doesn't fit is ignored, unless you can get a publication out of it. Love isn't a concept that would even come to mind concerning the natural world. The natural world is just a means to an end. A thing to be dissected, so they can get tenure. I was talking to a local botany professor, about how geology can drive speciation/change, and he was actually surprised to consider anything outside of genetic mechanisms. I was surprised at his surprise: his view just seemed so limited. A plant to him is an isolated, discrete entity, rather than the expression of the complex interactions and relationships between all the entities/factors in the environment going back 3.5 billion years."

❀ ❀ ❀

Or there's this. I just saw a snuff video of scientists pouring molten aluminum into an anthill to reveal the shape of the tunnels. Then the scientists marveled at the beauty of the shape of the anthill they just massacred to the last ant.

This is human supremacism.

❀ ❀ ❀

Or there's this. The air around the world has recently been declared to be as carcinogenic as second hand smoke. The leading cause of lung cancer is now industrial pollution.

This is human supremacism.

❀ ❀ ❀

Or there's this. Recently some people traveled by ship from Japan to Australia. Along the way, they saw plenty of garbage, but they saw no sea life, save one diseased whale.

This is human supremacism.

❋ ❋ ❋

Or there's the news article I just read that begins, "Ocean acidification due to excessive release of carbon dioxide into the atmosphere is threatening to produce large-scale changes to the marine ecosystem affecting all levels of the food chain, a University of BC marine biologist warned Friday. Chris Harley, associate professor in the department of zoology, warned that ocean acidification also carries serious financial implications by making it more difficult for species such as oysters, clams, and sea urchins to build shells and skeletons from calcium carbonate. Acidic water is expected to result in thinner, slower-growing shells, and reduced abundance. Larvae can be especially vulnerable to acidity. 'The aquaculture industry is deeply concerned,' Harley said. 'They are trying to find out, basically, how they can avoid going out of business.'"[4]

This news article follows the pattern of nearly every other news article about the murder of the planet, in that it jumps immediately to what most people in this culture are most concerned about: how this atrocity will affect the economy. Here's another headline: "Revealed: How Global Warming is Changing Scotland's Marine Life." The first sentence: "Global warming could cut commercial fish catches around Scotland by 20% while they increase by 10% around the south of England."[5]

Or how about this? Headline: "Climate Change Will Starve the Deep Sea, Study Finds." The article begins: "It's a vast, frigid abyss, where light rarely penetrates, and oxygen is in short supply. Its very otherworldliness has helped it seep into cultural awareness through science fiction and horror stories, but for most people the deep sea barely seems like a real place, let alone an important one. That's why the news this week that climate change is expected to lead to staggering losses in deep-sea life may not have seemed nearly as relevant as the traffic report or weather forecast. Whether or not it's public knowledge, however, the deep sea is home to thousands of commercially important species and is one of the last frontiers for new species discovery. The creatures of the deep are also key to the cycling of nitrogen, carbon and silicon in the ocean, a process that maintains the delicate balance of ocean life."[6]

So, the oceans are being murdered. The unimaginably complex, beautiful, once-fecund oceans—home to the majority of life on this water planet—are being murdered, and these articles quickly begin discussing how this will affect the industries that rely on exploiting the oceans?

This is human supremacism.

This is obscene.

This is routine.

This is why the world is being murdered.

✻ ✻ ✻

CBS News headline: "Salt-Water Fish Extinction Seen by 2048." Terrible news for the entire planet, right? Well, we all know what's really important; one of the researchers is quoted as saying, "If biodiversity continues to decline, the marine environment will not be able to sustain our way of life."[7]

Gosh, the real tragedy of the murder of the planet is that if the planet is dead, it will no longer be able to support our way of life.

✻ ✻ ✻

I hate this fucking culture.

✻ ✻ ✻

And these scientists do understand that it is this way of life that is killing the planet, right? This way of life that is dependent upon theft from all other communities, right? "If biodiversity continues to decline, the marine environment will not be able to sustain our way of life."

I don't understand how members of a species who considers itself the smartest on the planet can say so many things that are so stupid.

And how can members of a species who considers itself the smartest on the planet *do* something so stupid? Is it possible to be more stupid than to destroy the planet we live on?

Oh, that's right, unquestioned beliefs are the real authorities of any culture. And if some of the beliefs we must not question include the notions that human communities who do not share our unquestioned beliefs and values are not real communities; and that nonhuman communities (who certainly don't share our unquestioned beliefs and values) are not real communities; and that theft from these (not real) communities is not theft; and that murder of these (not real) communities is not murder or genocide; and that our (not real) theft and (not real) murder of these other (not real) communities can continue forever; that the point of existence is to commit these (not real) thefts and murders; that these thefts and murders will not severely impinge upon our ability to steal from and murder these others; that one of the most unquestioned beliefs in our culture must be that we must never question our inability or unwillingness to question these beliefs; and that the real pity of a murdered planet is that we can no longer continue to steal from or murder it, then I guess we can understand how someone can say something so absurd.

❃ ❃ ❃

So, I say to you, "I've heard that every member of your family is dying from poison released by the factory I put in next door. You look pretty sick yourself. I'll bet you aren't going to last another month. And your kids? That's one of your children over there? Is she still alive? She looks . . . this is just awful. Unbelievable. I'm so sorry to hear that you're all dying, because if you're all dead, who's going to eat at my restaurant, buy shoes at my shoe store? Who's going to work in my factory? This is all going to drive me out of business . . ."

❃ ❃ ❃

Just so you know my paragraph above isn't hyperbolic, here is a headline from today: "Mussels Could Soon be Off the Menu: Climate Change May Wipe Out the Shellfish if Acid in Oceans Stops Their Shells Forming." The first sentence: "The days of ordering delicacies such as moules marinières could be numbered, as climate change threatens to change the acidity of oceans."[8]

The other articles at least briefly mentioned the murder of the planet before quickly shifting their attention to what is clearly most important to them: the economy. But this article gives up even the pretense of caring about the planet and gets to the real point: how will this affect my access to delicacies?

This is human supremacism.

※ ※ ※

I just read the following description of sociopaths: "Imagine, if you can, feeling absolutely no concern for another ~~human~~ being. No guilt. No remorse. No shame. Never once regretting a single selfish, lazy, cruel, unethical or immoral action in your entire life.

"Nobody matters except you. Nobody deserves respect. Equality. Fairness. They are useless, ignorant, gullible fools, who are taking up space and the air you breathe.

"Now I want you to add to this strange fantasy the ability to conceal from other people exactly what you are, to be able to hide your true nature. Nobody knows what you're really like . . . how little you care for other people [including nonhuman people] . . . what you're capable of . . .

"Imagine what you could achieve. Where others hesitate, you will act. Where others set boundaries, you will cross them, unhampered by any moral restraints or pangs of disquiet, any rules or ethics, with ice water in your veins and a heart of pure stone."[9]

This is a description of sociopathy.

This is a description of human supremacism.

CHAPTER ONE

THE GREAT CHAIN OF BEING

The Courtier disdaineth the citizen;
The citizen the countryman;
the shoemaker the cobbler.
But unfortunate is the man who does not have anyone he can
look down upon.

TOMAS NASH, 1593

What really fascinated him was. . . . [P]ossessing them physically
as one would possess a potted plant. . . . Owning, as it were, this
individual.

SERIAL SEX KILLER TED BUNDY

One of the most harmful notions of Western Civilization—and one of the most foundational—is that of the Great Chain of Being, or Latin *scala naturae* (which literally means 'ladder or stairway of nature'), closely related to the divine right of kings. It is a hierarchy of perfection, with God at the top, then angels, then kings, then priests, then men, then women, then mammals, then birds, and so on, through plants, then precious gems, then other rocks, then sand. It's a profoundly body-hating notion, as, according to those who articulated the hierarchy, those at the top—the perfect—are pure spirit; and those at the bottom—the imperfect, the corrupt—are pure matter, pure body. Then both men and women live in a battleground of spirit and body, with men tending to be put more in the box representing mind/spirit/better/perfected, and women tending to be put more in the box representing body/life/death/corruption/imperfection. In this construct, humans are the center of attention, with those above humans being bodiless and perfected, and those below being fully embodied, imperfect, and having no mind. Of course, within each category there are sub-categories. So civilized man is far more perfected than 'primitive' man, who is barely removed from animals. You see this hierarchy everywhere within this culture, only now as we've secularized we've gotten rid of God and angels, leaving civilized (especially white) men at the top.[10] And of course, those at the top get to use those below however they want. For example, men have access to the bodies of women, because men are higher on the hierarchy than women.

The Great Chain of Being has long been used to rationalize whatever hierarchies those in power wish to rationalize. It has been and is central to the notion of the Divine Right of Kings, to racism, to patriarchy, to empire. It is a very versatile tool.

The Great Chain of Being also underlies the modern belief that the world consists of resources to be exploited by humans. Traditional Indigenous peoples across the earth do not believe in this hierarchy; instead, they believe the world consists of other beings with whom we should enter into respectful relationship, not inferior others to be exploited. This is one reason these other cultures have often been sustainable.

Our perception of evolution is infected with this belief in the Great Chain of Being, as so often people, including scientists, think and write and act as though all of evolution was about creating more and more perfect creatures, leading eventually to that most perfect creature yet: us.[11]

* * *

Did you know that mother pigs sing to their children?

And pigs dream.

And pigs have a good sense of direction, and can find their way home from great distances. They learn from watching each other. And they will outsmart each other: one pig will often follow another to food before grabbing it away; the other pig will then change her behavior so she won't get fooled again (which is more than we can often say for many humans).

Scientists have done experiments where they trained pigs to use their snouts to move cursors on video screens. They found the pigs could distinguish between (human) scribbles they had seen before and (human) scribbles they had not. Pigs learn this skill as quickly as do chimpanzees.

Pigs are capable of abstract representation. They can hold an icon in mind, then remember it till a later date. They can also remember verbal commands—and these commands are given in a human language; I'd like to ask how many words of pig you or I know—and when these commands are repeated several years later, they will still remember what to do.

❀ ❀ ❀

I need to go into town this afternoon, so this morning I made a list to remind myself where I need to stop. Evidently, I can't remember my own instructions for even several hours. And who's the smart one?

❀ ❀ ❀

Pigs form complex relationships with their peers. They have friends. Of course. How could anyone think otherwise? They sometimes work in pairs to open gates and will open other gates to release other pigs.[12]

❀ ❀ ❀

I'll tell you the image I can't get out of my head. It's of a mother pig confined to a tiny crate, suckling her children. And singing. To her children. Human supremacists have stolen her freedom, but they've not been able to steal her capacity to love.

And humans—the ones who put her in the cage—are superior?

❀ ❀ ❀

Here's another image I can't get out of my head. It's of a mother dolphin singing to her child. But they are both dying.

Here's why.

Mother dolphins nurse their young for eighteen months, longer than many humans. The mother dolphins love their children with fierceness and loyalty. Even when a baby dolphin is caught in a tuna net, the mother will often not abandon the child, but move in close, and comfort and sing to her baby until both are drowned in the net.

Fishing companies acknowledge that most of the dolphins they kill are children and their mothers, who will not leave them even unto death.

❀ ❀ ❀

And did you know that cats can count? A litter of kittens was caught in a house fire. Their mother kept returning to the fire to bring out her children, one by one. The fire and smoke blinded her. When the kittens were all safe outside she made sure she had them all by touching each one with her nose, counting.

She did not regain her eyesight. But while she took care of her kittens as they grew, she would always make sure to touch each one with her nose, to make certain they were all there.

❀ ❀ ❀

Human supremacists could argue that she was not counting them, but instead smelling them.

I would argue these human supremacists are missing the point entirely.

❀ ❀ ❀

Sometimes, because I eat meat (that doesn't come from factory farms), vegans have accused me of speciesism. But the truth is quite the opposite. I don't believe in the Great Chain of Being. I believe that plants are every bit as sentient as anyone else. Human supremacists draw the line of being/not-being between humans and nonhumans, with humans being sentient and having lives worth moral (and other) consideration, and nonhuman animals, not so much. Vegans often draw the line of being/not-being between nonhuman animals and plants, with nonhuman animals being (to varying degrees) sentient and having lives worth moral (and other) consideration, and plants, not so much. I don't see it that way. I believe that no matter whom you eat, you are eating someone.

Not infrequently, vegans have, seemingly without a sense of the irony of doing so, mocked my and many others' beliefs in plant sentience using the same scorn, and indeed sometimes the same words, with which many meat-eaters dismiss vegan beliefs in nonhuman animal sentience. Same scorn, same uncrossable line between meaningful subject and meaning-less object; the line is merely drawn at a different place.

I try not to draw the line at all. When I was a child I didn't draw it. I don't think most of us do. And then I was, like nearly all of us, taught to draw that line, by religion, by science, by entertainment, by art, by day-to-day interactions with those who had themselves been taught to draw that line.

I remember twenty-some years ago I interviewed the great environ-mental philosopher Neil Evernden. He said he didn't believe in drawing a line between meaningful humans and non-meaningful nonhumans. I was confused, and curious, and asked him where, then, *would* he draw the line between those who are meaningful and/or sentient, and those who are not? I'll never forget the liberation—the homecoming—I felt when he asked, sincerely yet clearly rhetorically, "Why do we need to draw that line at all?"

❀ ❀ ❀

Once that question is asked, meaning that the assumptions of human supremacism, and more broadly the Great Chain of Being, have begun to be questioned, the supremacists'—and it doesn't much matter whether we're talking about male, white, civilized, or human supremacists—next line of defense is often religion. We are superior because God says we are. It's a crude yet powerful defense of any supremacism because no one can prove you wrong. It's effectively a conversation stopper. Like the bumper sticker tells us: "God says it, I believe it, that settles it." No thought required (or allowed). Nothing to dispute here. Move along.

So a couple of thousand years ago some people told themselves and their friends the rather flattering story that they alone among species were created in God's image (and that their culture was God's culture, and that their sex was God's sex), and that God told them to tell everyone else that God gave them dominion over the earth. The followers of these stories have pillaged their way across the planet, bringing us to today, when all major US Presidential candidates call themselves believers in and followers of this story, and all of them are in their respective capacities presiding over what amounts to the final plundering and murder of the planet. But God gave Man dominion, didn't He?

As I said, it's crude, but effective as a defense. It says nothing about whether this human supremacism is in any way warranted. I was going to add that it also says nothing about whether this supremacism is moral, but of course it does, by declaring it supremely so, which is precisely the point and *raison d'être* of any (in)decent supremacist religion.

This brings us to what is often the next line of defense for different types of supremacism: science. The point here is not to discuss how science has been used to support slavery, genocide, gynocide, ecocide, white supremacism, male supremacism, and on and on. I've described this in book after book. The point here is also not to discuss how science is founded on and implicitly fosters human supremacism in that it is based on the notion of the intelligent, meaningful, superior, spectating, and speculating human subject who observes, measures, and controls (bends to his will, or, as scientific philosopher Richard Dawkins puts it, makes jump through hoops on command) a less- or unintelligent, meaningless, inferior, acted-upon object or Other: a resource. This is a subject I have also treated in book after book. The point here is not to discuss how science has from the beginning been about increasing human control over the rest of the world. The point here is not to discuss the great harm science (both as a worldview and in its applied forms) has done to the real world. Once again, I have discussed in book after book ways in which the real world is not better off because of the existence of science as a field of study. The point here is not to discuss how science is based on the same old Great Chain of Being, with Perfect and Abstract Reason (Scientific Law, Abstract Law, Abstract Economics, and so on) now substituted for a Perfect and Abstract Distant Sky God, and with machines now substituted for angels, since of course machines are far more perfect than humans, far more controllable and not subject to "human error." They also do not die. These are just a few of the ways science supports human supremacism. Remember, unquestioned beliefs are the real authorities of any culture.

The point I'm more interested in emphasizing here is that most of the scientific (and "common sense") arguments used to defend human supremacism (and the same is true for various scientific and "common sense" means that have been used to defend white or male supremacism) are tautological, in that humans are using themselves as the standard by which all others are judged. Here's another way to say this: humans choose human characteristics as the measure of what characteristics define superiority. It doesn't much matter whether you're a member of a religious group that decides you are the Chosen People and says that some God only you can hear told you that you and you alone are made in the image of this omnipotent God; or whether you're more modern and project an anthropocentric version of sentience onto the real world, whereby beings are considered sentient primarily based on how closely they resemble humans. In either case you're projecting this culture's destructive notion of a hierarchical Great Chain of Being onto a beautiful, vibrant, living, sentient world full of others.

That's fucking nuts. Or convenient. Or both.

Let me give an example.

Or maybe not. Honestly, after more than twenty books of talking about this bullshit, I'm sick of it. Let's cut to the chase. I'm superior to you. I'm smarter than you. I'm more sophisticated than you. My life is more meaningful than yours. You are, frankly, insignificant compared to me. I am a sentient being. You are a resource for me to exploit. It's that simple. How do I know this? Because I have more than twenty books out and you do not. Having more than twenty books out is the measure by which I have determined that we judge superiority, intelligence, sophistication, and a meaningful life. How do I know that's the measure? Well, I decided that's the measure. And besides, it's just common sense. I remember years ago I read (or maybe someone told me, or maybe I made it up) that the most complex logical task anyone can do is write a book. It doesn't actually matter to me whether it's true, because it sounds right, and also because it supports my superiority.

Look at it this way: obviously, I am able to think, and obviously I am able to communicate. By definition. Or else I wouldn't have more than twenty books out. And just as obviously, you are not able to think, and you are not able to communicate. Because if you did, you, too, would have more than twenty books out! QED. I know what you're going to say: *Stephen King*. How can I be superior when Stephen King has written, at last count, more books than it is possible to count? But that's easy: I'm still superior because most of his books are novels. I understand that some of my books are novels, too, but mine still count and his don't because mine are strongly pro-feminist novels about fighting back against those who would abuse women or the land, which means they and I are both superior. Yes, I know King wrote *Dolores Claiborne*, but that book doesn't count, because, well, I'm sure you can see why it doesn't count, right? Same with *Rose Madder*, by the way. I'm still superior and smarter. Aren't I?

And don't even talk to me about all the other things you might have done. It doesn't matter. If you don't have more than twenty books out, I'm superior to you, smarter and more significant than you are. What? You say you raised two children to be happy and healthy adults? Well, first I would correct your language. You didn't "raise children"; you "produced offspring." And clearly producing offspring can't be a sign of intelligence or superiority; for crying out loud, mice produce offspring every six weeks. Producing offspring is not like writing, because writing is truly creative. It is the superior mind creating something out of nothing, with no help from anyone else. Producing offspring is just instinct, and takes no talent or creativity whatsoever: a female comes in heat, is mounted by (never makes love with) a male (while being watched through binoculars by David Attenborough, who already a bit-too-breathlessly described their courtship rituals), and their genes are passed on. Nothing but hormones and instinct. And same with the so-called raising of these so-called children. Instinct, instinct, instinct. Just like it is with so-called mother bears and so-called mother elephants (and while we're at it, let's call them mother trees and mother bacteria (LOL)). For crying out loud, if a "mother" mouse can "raise" her "babies," where's the talent? Where's the exclusivity? Birds do it. Bees do it. What makes *you* feel so special? So-called *procreation* (snort) is sure as hell not as much of a miracle as a book, which is a wondrous creation of the (or especially my) mind.

Now that we've so clearly shown we can dismiss mere bio "creation" as any sort of sign of intelligence or superiority—it's just instinct and "natural"—let's move on to other so-called creations. What about monuments or other "great" engineering "achievements"? Why do my books qualify me as smarter than and superior to, for example, an engineer who "creates" a dam? Doesn't it take intelligence and superiority to build a dam? Well, apart from the fact that I—not dam builders—am defining the qualifying characteristics, I've only got one word for you: beavers. Seriously, beavers build dams, and what are beavers? They're nothing but rodents with big teeth. Instinct! (Never mind that beavers teach their children how to make dams.) And besides, beaver dams make some of the most biodiverse habitat in the world, and engineer-made dams kill rivers. So how does it feel, Mr. Big Shot Engineer, to be less competent than a fucking rodent with big teeth? Once again, where's the talent?

Another reason I'm superior to you is that I have a Bachelor's of Science degree in Mineral Engineering Physics. That proves I'm smart, in fact smarter than anyone who has ever lived, including those who died long before there was such a field of study as physics, and so never had the opportunity to get a degree in it. If they were so smart, they would have figured out a way. Likewise, my superiority based on my writing books extends over those who lived before the invention of the printing press; if they were as high as I on the Great Chain of Being, they would have overcome this trivial obstacle. And once again, I know what you're thinking: given the superiority of someone with a BS in physics (with an emphasis in BS), why, then, are those with a Master's or a PhD not even more superior? I think that's pretty obvious: I was smart enough to not stay in physics.

And here's another reason I'm smarter than and superior to you, and that my life is meaningful while yours is not: the color of my skin. It's white, or more precisely, if you don't mind just a tiny bit of completely-deserved arrogance, my skin color is flesh tone. Why does my white (flesh-tone) skin make me superior to you? Because I'm white, that's why.

Yet another reason I'm smarter than and superior to you has to do with my chest hair. I have just the right amount, which is some but not a lot. If I had a lot, of course that would make me too much like an animal, which would make me inferior. And if I had none, then I would make me too much like a woman, which would also make me inferior.

And how else do I know I'm smarter than you, that my life is significant and yours is not? Because I have a penis. I'm a man. A Man. I know what all you other men are thinking: you're thinking that you have a bigger, better penis than I do. But let me assure you: you don't have mine. I have it on the very best authority, in fact the only authority that matters—mine—that my penis is special.

❀ ❀ ❀

Just to be clear, if you don't understand or don't agree with what I wrote above, well, that's just another sign of your inferiority.

❀ ❀ ❀

So here's a question: if you can easily see the offensiveness, arrogance, stupidity, falsity, and tautology of the above arguments when they apply to me declaring myself smarter than and/or superior to you, why do so few people see the same when it comes to humans declaring themselves smarter than and/or superior to nonhumans?

❀ ❀ ❀

None of which alters the fact that humans *are* smarter than *and* superior to all others because we've got really big brains. As one rather narcissistic website proclaims, "The human brain is the most complex phenomena [sic] in the known universe."[13]

Really? The human brain is more complex than oceans? Than forests? Than the sun? This culture likes its narcissism undiluted. But what else would we expect from a culture whose members designate themselves *Homo sapiens sapiens*: the wisest of the wise?

There are at least four problems with the notion that humans are smarter and superior because of the size of our brains.

The first is that it's the same old tautology. In math, a tautology is "a logical statement in which the conclusion is equivalent to the premise." You could also call it circular logic. It's like the old joke, "The first rule of the tautology club, is the first rule of the tautology club." Here is the tautology. Humans have big brains. Humans decide big brains are a sign of intelligence and/or superiority. Therefore, because humans have big brains, humans must be more intelligent and/or superior. Now, that was a surprise, wasn't it?

The second problem, and we'll discuss this more later, has to do with whether it is meaningful or appropriate to attempt to make intelligence comparisons across species. Humans are more intelligent at what? And so far as this larger exploration—human supremacism—humans are superior at what?

The third problem is that I'm not convinced we think only with our brains. I think we think partly with our brains, but also with our whole bodies, and with our surroundings. If creatures think only with brains, how, for example, do wasps, with brains only one-millionth the size of human brains, correctly differentiate between photographs of the faces of other wasps? I thought only "higher" animals, like, uh, the highest of the high animals, humans, were able to differentiate between photos of their friends. And how do honeybees know how to communicate through dance? As one writer states, "Honeybees don't have much in the way of brains. Their inch-long bodies hold at most a few million neurons. Yet with such meager [sic] mental machinery [sic] honeybees sustain one of the most intricate and explicit languages in the animal kingdom [sic]. In the darkness of the hive, bees manage to communicate the precise direction and distance of a newfound food source, and they do it all in the choreography of a dance. Scientists have known of the bee's dance language for more than 70 years, and they have assembled a remarkably complete dictionary of its terms, but one fundamental question has stubbornly remained unanswered: How do they do it? How do these simple [sic] animals encode so much detailed information in such a varied language?"[14]

My fourth problem with the notion that humans are smarter than and superior to all others because we've got really big brains is that if we *do* think with our brains, we have to recognize that there are plenty of beings with brains larger than ours. Hell, human beings 5,000 years ago had brains 10 percent larger than those of human beings today. I guess that means we're 10 percent stupider than we were back then. This would explain many things, from fundamentalisms of all stripes to pop culture to presidential politics to environmental policy to the existence of bright green environmentalism to the stupid arguments of human supremacists. Neanderthals also had larger brains than do humans, which might explain why Neanderthals never created insurance advertisements claiming that something is so simple even a caveman can do it. In any case, if intelligence or superiority is measured by brain size, humans lose. Average human brain weight is somewhere around three pounds. Walruses are nearly as big, at almost two and a half pounds. Elephants are much larger, at more than ten pounds. And whales run from four and a half pounds up to more than seventeen pounds.

But wait, I can hear you say, changing the rules as we go, *actual brain size isn't important. Brain size to body mass ratio is what leads to intelligence, and specialness, and all sorts of scrumptious wonderfulness that makes humans meaningful and everyone else meaningless! Of course a big animal needs a big brain to control its movements. Or something.*

Let's leave off the fact that big animals don't really need big brains—the Stegosaurus weighed four or five tons and had a brain weighing less than three ounces—and take this one at face value. The human brain is about 2.5 percent of our body weight. Sadly for us, this is about the same as it is for mice. The brains of small birds make up about 8 percent of their body weight. The brains of shrews are about 10 percent of their body weight.

Well, that's embarrassing. I guess since we didn't win either of those contests, we'll have to come up with another way to determine intelligence. As one author puts it, while arguing that humans are unquestionably more intelligent than anyone else: "Neither absolute brain weight nor the relationship between brain weight and body size provide us with sensible criteria for comparing the intelligence of different species."[15] Of course they don't; any sensible criteria would make it so we're number one.

We could have predicted that supremacists would be quick to propose lots of variants on this same human supremacist theme: for example, that intelligence is determined by the quantity brain size minus spinal column over body weight, and on and on.

It doesn't really matter, so long as we win.

But the whole "humans are smarter/superior because of brain size" theory has a bigger problem than either whales or shrews. The bigger problem is fungi. As I wrote in *Dreams*, "Did you know that fungi are intelligent? I didn't. But [Paul] Stamets writes [in *Mycelium Running*], 'I believe that the mycelium operates on a level of complexity that far exceeds the computational powers of our most advanced supercomputers.' And he backs this up. Fungi demonstrate simple straightforward intelligence (even measured by our own narcissistic standards); if you put a slime mold[16] at one end of a maze, it will grow randomly until it finds food. If you take a piece of this slime mold and put it in the same maze, it will remember where the food is, and grow directly toward it, with no false turns. Further, if you compare the information-transferring organization of mycelium to the organization of the Internet, you'll find that, as Stamets says, the 'mycelium conforms to the same mathematical optimization curves that Internet theorists and scientists have developed to optimize the computer Internet.' Or rather, the Internet conforms to the same curves as the mycelium." Fungi can be seen as huge neurological nets. Back to *Dreams*, where I begin by citing Stamets, "'I believe that mycelium is the neurological network of nature. Interlacing mosaics of mycelium infuse habitats with information-sharing membranes. These membranes are aware, react to change, and collectively have the long-term health of the host environment in mind.'

"I had to read that last phrase three times. But now I get it.

"Have you ever wondered, for example, how tiny trees survive in the shade of their much-larger elders?

"I asked Paul Stamets about this.

"He answered, 'If you've been in an old-growth forest, you've probably seen hemlock trees on rotting nurse logs. They're usually the first trees to come up in these highly shaded environments. They have very little exposure to light. When these small saplings were dug up and taken to a greenhouse-like environment and given a similar amount of low light, they all died. The question became, where are these young trees getting their nutrients? So researchers radioactively tagged carbon and watched the translocation of carbon in the forest. They found that birch and alder trees growing in more riparian habitats along rivers—where there is more sunlight—were contributing nutrients to the hemlocks via mycelium.'

"I said, 'Wait. Are they . . .'

"He said, 'Yes. The mycelium—which, if you remember, run all through the forest soil and connect different parts of the forest to each other—are transferring nutrients from trees of one species who have nutrients to spare to trees of another species who need nutrients or they will die. The mycelium is taking care of the health of the forest. I think they are doing this because they understand that the health of the biodiversity of the ecosystem speaks to the survival of these fungal populations. I think these fungi have come to learn through evolution that biodiversity and resiliency of ecosystems are for the benefit of all the members, not just for one.'"

Above, I said that fungi can be seen as huge neurological networks. Let's step away from the scientific language and say *brains*. Now let's talk about how huge these brains can be. A final time from *Dreams*: "Fungi can grow to be immense. A single *Armillaria ostoyae*—or honey mushroom—growing in Oregon covers 2,200 acres, or more than three square miles. It's estimated to be 2,400 years old, and weighs more than six hundred tons. For hundreds and then thousands of years, this fungus has lived below and at the surface of the soil. It has nurtured tree upon tree, forest upon forest. It has attended to the needs of these trees, the needs of this forest. It has fed them. It has witnessed fire and rain, snow and drought. It has been parent and child to this forest; lover and friend; killer, decomposer, and creator. And through all of this it has built up the soil upon which all depend. Unfortunately, now it has also witnessed chain saws and clear-cuts. It has witnessed logging-induced destruction of soil. It has suffered herbicides and fungicides. It has suffered logging roads that have cut it into pieces. It may very well be witnessing the end of the forest, the forest it has lived with and loved so well."

Nearly all of the mass of this fungus consists of neurological network, so I guess that throws both our "brain size" and "brain-to-body-mass ratio" criteria out the window.

CHAPTER TWO

LANGUAGE

People don't think of trees as alive. We never see them moving unless the wind disturbs them, and then it is not their movement but the wind's. The Wart saw now that trees are living, and do move. He saw all the forest, like the sea weed on the ocean's floor, how the branches rose and groped about and waved, how they panted forth their leaves like breathing (and indeed they were breathing) and, what is still more extraordinary, how they talked.

If you should be at a cinema when the talking apparatus breaks down you may have the experience of hearing it start again too slowly. Then you will hear the words which would be real words at a proper speed now droning out unintelligibly in long roars and sighs, which give no meaning to the human brain. The same thing happens with a gramophone whose disc is not revolving fast.

So it is with humans. We cannot hear the trees talking, except as a vague noise roaring and hushing which we attribute to the wind in the leaves, because they talk too slowly for us. These noises are really the syllables and vowels of the trees.

T.H. WHITE

Humans are unique in their capacity to have language. It seems like everyone within the dominant cultural tradition, from linguists to men and women on the street, says this. But this is all crap. Prairie dogs have language and grammar. Chickens have language. Cows and sheep and goats have dialects. Elephants have language. Whales have language. Dolphins call each other by names they've made up. Orcas can learn to speak the language of bottlenose dolphins. And seriously, when bonobos have learned how to translate between humans and other bonobos and tell humans what the bonobos are saying, and vice-versa, who is the smart one? How many humans can speak bonobo? Likewise, scientists have debated for years whether the nonhuman apes who are kidnapped and taught American Sign Language really have language or are merely mimicking (and how, besides mimicking, do they think human children (and adults) learn language?).[17] But the scientists consistently refuse to ask these key questions: 1) The nonhuman ape knows its own language plus human ASL; which of these creatures is bilingual, knowing the language of another *species*? 2) How could you possibly learn another's language if you do not believe this other has language? 3) If a bunch of nonhuman apes in white lab coats kidnapped a scientist and put him in a nonhuman ape version of a mobile home, how long would it take for the scientist to learn their language? 4) Why do scientists think they have the right to kidnap a nonhuman ape? and 5) If we judge these nonhuman apes and these scientists by whether or not they kidnap innocent individuals, which species is morally superior?

Likewise, there's a border collie who knows more than a thousand words. Of English. Human. The dog also has a sense of syntax. Seriously, how many words of border collie do you know? If you're like most of us, I'm guessing that would be one: woof. After which many would say to the dog, "Quit barking! I'm trying to watch television!"

And of course, bees have their language of dance.

And lots of other species speak. Perhaps they all do. How would we know? We're all too busy trying to rationalize how their languages aren't really languages, how, as Europeans said about the human Africans the Europeans were enslaving, "when they speak they fart with their tongues in their mouths."

I've often wondered what trees think about our perhaps to them rudimentary language . . .

Wait . . . what? Back up. First, trees think? That's crazy!

Do you remember a few pages ago where I mentioned that sometimes vegans dismiss the notion of plant intelligence in the same scornful way that some meat-eaters dismiss the concerns of vegans? I wasn't making that up. And of course it's not only vegans who do that, but many people in this culture.

Vegans provide an interesting case because many at least say they oppose human supremacism. But, of course, their analysis is predicated on the Great Chain of Being. A couple of years ago I read an essay in which a vegan attacked a belief in plant sentience by calling it a "bizarre and unsupportable" claim "held by not a single reputable scientist in the world." First, let's explicitly point out that members of the National Cattlemen's Association would be quick to use similar language to dismiss nonhuman animal sentience. Next, let's in this moment not take too much issue with the absurd, racist, and counterexperiential notion that, once again following the Great Chain of Being, scientists have a monopoly on understanding reality. We'll only take issue long enough to ask why we should believe that scientists have a better understanding of whether or not plants are sentient than do Indigenous peoples who have formed millennia-long fully mutual relationships with these plant communities, or for that matter, a better understanding than anyone who forms long-term personal relationships with any plant species (I'm thinking, to provide one example among many, of long-term marijuana growers who recognize the personalities of their individual plants). Having said that, we'll move on to take this notion at face value.

Let's say hello to Stefano Mancuso, one of the many reputable scientists who argue that plants are sentient. Mancuso is director of the International Laboratory of Plant Neurobiology in Florence, Italy. What is plant neurobiology? Their website states that it is a field where "all the plant sciences will meet together to study diverse aspects of signaling and communication at all levels of plant organization, starting from single molecules and ending at ecological communities. Twentieth-century biology was dominated by attempts to reduce extremely complex biological phenomena to the actions of single molecules. While this process will continue in future, we also need to integrate the avalanche of obtained data together using system-based approaches. Plant Neurobiology will cover all plant sciences under one umbrella from the perspective of signaling and communication at all levels of biological organization. Plant Neurobiology will interlink together molecular biology with physiology, and behavior of individual organisms, up to the system analysis of whole plant societies and ecosystems. This integrative view will allow our understanding of communicative plants in their whole complexity."

I note with delight their use of the phrase "plant societies."

I also note with delight that Mancuso has videos showing plants doing what he describes as "playing." Watch the videos for yourself and see if you agree.[18]

The Laboratory website continues, "Our viewing of plants is changing dramatically away from passive entities being merely subject to environmental forces and organisms that are designed solely for accumulation of photosynthate. In contrast, plants emerge as dynamic and highly sensitive organisms that actively and competitively forage for limited resources, both above and below ground, organisms that accurately compute their circumstances, use sophisticated cost benefit analysis, and that take defined actions to mitigate and control diverse environmental insults. Moreover, plants are also capable of a refined recognition of self and non-self and are territorial in behavior. This new view sees plants as information-processing organisms with complex communication throughout the individual plant. Plants are as sophisticated in behavior as animals but their potential has been masked because it operates on time scales many orders of magnitude less than that operating in animals."

I would quibble with their use of machine language, such as "systems" and "computing," but I'd be also clear this is only a quibble. In my own mind I'll just perform internal substitutions like *thinking* for *computing* and happily move on.

The website continues, "Due to this lifestyle [of operating on slower time scales], the only alternative to rapidly changing environment is rapid adaptation [because they can't run away]. Therefore, plants have developed a very robust signaling [i.e., communicating] apparatus. Signaling in plants encompasses both chemical and physical communication pathways. The chemical communication is based either on vesicular trafficking pathways, as accomplished also across neuronal synapses in brains, or through direct cell-cell communication via cell-cell channels known as plasmodesmata. Moreover, there are numerous signal molecules generated within cell walls and also diffusible signals, such as NO, ROS and ethylene, penetrating cells from exocellular space. On the other hand, physical communication is based on electrical, hydraulic, and mechanical signals [as is ours]. Besides interaction with the environment, plants interact with other communicative systems [i.e., beings] such as other plants, fungi, nematodes, bacteria, viruses, insects, and predatory animals."[19]

"Interact with other communicative systems." In a non human-supremacist culture, we would all be comfortable just saying they talk to each other.

And they do. For example, plants tell other plants that herbivorous insects are eating them, and the plants who receive this message prepare defenses against the insects. After all, the plants can't run away, so they need to have some way to keep from being eaten. One way plants communicate is by releasing pheromones that tell other plants to prepare. They also release pheromones calling predator insects. I've witnessed the effects of this, as a tree being devastated by an overwhelming aphid infestation was saved by the timely arrival of thousands of lady bugs, and soon, their young. But it gets even better. Plants can *hear*, and they respond to what they hear (which is more than can be said for global warming denialists, and frankly a lot of others). They can hear the sounds of leaves being eaten by caterpillars, and respond by changing the composition of their leaves to make them less palatable.[20]

And it gets even better. Scientists have covered plants with plastic bags containing herbivorous insects, and the plants have *still* been able to communicate to their relatives, friends, and neighbors that they need to prepare their defenses. In this case, how? Through the mycelial networks. The fungi facilitate their communication. More on this in a moment.

It's a good thing no reputable scientists believe in plant sentience.

Now let's say hello to Charles Darwin, who also wrote on the intelligence of plants. In 1880 he published a book called *The Power of Movement in Plants*," in which he wrote, "It is hardly an exaggeration to say that the tip of the radicle . . . acts like the brain of one of the lower [sic] animals; the brain being situated within the anterior end of the body, receiving impressions from the sense-organs, and directing the several movements."[21]

It's a *really* good thing no reputable scientists believe in plant sentience.

If you Google "plant intelligence" you can see that articles on plant intelligence have been published in dozens of even mainstream sources, including *The New York Times*, the *Christian Science Monitor*, the *Guardian*, and for crying out loud, *Wired*. *The New York Times* article was entitled, "Sorry, Vegans: Brussels Sprouts Like to Live, Too." As the author notes, "The more that scientists learn about the complexity of plants—their keen sensitivity to the environment, the speed with which they react to changes in the environment, and the extraordinary number of tricks that plants will rally to fight off attackers and solicit help from afar—the more impressed researchers become, and the less easily we can dismiss plants as so much fiberfill backdrop, passive sunlight collectors on which deer, antelope and vegans can conveniently graze. It's time for a green revolution, a reseeding of our stubborn animal minds."[22]

Or consider the following from a *New York Times* blog: "A team of scientists from the Blaustein Institute for Desert Research at Ben-Gurion University in Israel published the results of its peer-reviewed research, revealing that a pea plant subjected to drought conditions communicated its stress to other such plants, with which[23] it shared its soil. In other words, through the roots [and our old friends the mycelia networks], it relayed to its neighbors the biochemical message about the onset of drought, prompting them to react as though they, too, were in a similar predicament. Curiously,[24] having received the signal, plants not directly affected by this particular environmental stress factor were better able to withstand adverse conditions when they actually occurred. This means that the recipients of biochemical communication could draw on their 'memories'[25]—information stored at the cellular level[26]—to activate appropriate defenses and adaptive responses when the need arose."[27]

Or we can talk about an op-ed written by Anthony Trewavas of the Institute of Molecular Plant Science in the journal *Trends in Plant Science* called "Green Plants as Intelligent Organisms" that "assesses whether plants have a capacity to solve problems and, therefore, could be classified as intelligent organisms. The complex molecular network that is found in every plant cell and underpins plant behaviour is described. The problems that many plants face and that need solution are briefly outlined, and some of the kinds of behaviour used to solve these problems are discussed. A simple way of comparing plant intelligence between two genotypes is illustrated and some of the objections raised against the idea of plant intelligence are considered but discarded. It is concluded that plants exhibit the simple forms of behaviour that neuroscientists describe as basic intelligence."

In other words, plants make decisions as to where they should grow more roots, and where they should grow fewer. If there are more nutrients, they grow more. Given a choice between growing roots into soil with the roots of other plants, or no roots of other plants, they will choose the latter. They make similar decisions with leaves, deciding where to grow and how to face leaves, and deciding when to abandon these leaves and let them fall. And plants predict the future, then make decisions based on these predictions. The article cites neuroscientists Peggy La Cerra and Roger Bingham as stating, "The *sine qua non* of behavioral intelligence systems is the capacity to predict the future; to model likely behavioral outcomes in the service of inclusive fitness," and then goes on to note that in "recurrent and novel environmental situations, cells, tissues and whole plants model specific future behaviours so that the energetic costs and risks do not exceed the benefits that adaptive, resilient, behaviour procures. Such modeling takes place on an adaptive representational network, an emergent property constructed from cell transduction and whole plant networks."

Before we continue, I want to mention something about their definition of intelligence. They write, "The *sine qua non* of behavioral intelligence systems is the capacity to predict the future; to model likely behavioral outcomes in the service of inclusive fitness." My grandfather had diabetes. It affected his life. My mother has diabetes. It affects her life. I have known the relationship between diabetes and ingesting carbohydrates since I was a child. This didn't stop me from drinking pop in high school, drinking milkshakes (homemade: a pint of ice cream, three bananas, a pint of strawberries, a pint of milk, dashes of nutmeg and cinnamon, and three raw eggs; serves one) through my twenties, eating ice cream (four scoops and a banana) through my forties, and eating my beloved potatoes (in any form: baked (six at a time), mashed, scalloped, fried, JoJos, you name it—even, I must admit, raw—and oh my god, how could I forget in the form of potato chips) up to the age of fifty.[28] And then, at fifty-two, I learned, what a surprise, that I was pre-diabetic. Despite the fact that I had sufficient information (and presumably motivation), did I "model likely behavioral outcomes in the service of inclusive fitness"? Clearly, not at all. Clearly, I was not behaving in a manner indicative of having any native intelligence.

Trewavas gives examples of intelligent plant behavior. "Branch and leaf polarity in canopy gaps have been observed eventually to align with the primary orientation of diffuse light, thus optimizing future resource capture. The internal decisions that resulted in the growth of some branches rather than others were found to be based on the speculatively expected future return of food resources rather than on an assessment of present environmental conditions."[29]

Read that last sentence again. The plant is basing its decision on where to grow branches—which is a long term project, and in the case of trees can take years—on its prediction of how sunlight will come through the overstory by the time the branch is able to grow to that spot.

These plants not only show themselves capable of making better decisions—or being better at "modeling likely behavioral outcomes in the service of inclusive fitness"—than I, they are also more observant than I am. I, too, live in a forest, and I'm not sure I could predict what the overstory might look like in any particular place a few years from now.

Trewavas continues, "The Mayapple (*Podophyllum peltatum*), a forest floor plant, also makes commitment decisions as to branching or flowering years ahead, using a multiplicity of current environmental information. Many temperate trees make decisions about flower numbers a year ahead." Plants can predict not only future sun, but future shade, and make decisions as to how and where to grow based on these predictions. He says, "The stilt palm (*Socratea exorrhiza*) 'walks' out of shade by differential growth of prop roots."[30] This was a bit much even for me, but I researched it further. Here's what I found. The base of the trunk can be up to a meter off the ground, and is supported by an open cone of spiny stilt roots. When the palm finds that it would be better off elsewhere, say if a tree falls and knocks it over or blocks its sunlight, or if it germinates too close to the parent tree, it sprouts new roots, each one higher than the old ones. These roots reach out in the direction the plant wishes to go, and eventually the plant "walks" to the desired spot, at which time the old roots and the trunk below the new roots rots.[31]

Plants also plan for the future when it comes to water. Trewavas writes, "When provided with water only once a year, young trees learn to predict when water will be provided in the future and synchronize their growth and metabolism with this period only."

The author paraphrases two other scientists (Seeley and Levien), who say, "It is not too much to say that a plant is capable of cognition in the same way that a human being is. The plant gathers and continually updates diverse information about its surroundings, combines this with internal information about its internal state (simple reasoning) and makes decisions that reconcile its well-being with its environment."

Near his close Trewavas cites another scientist, who has studied the intelligence of predatory protozoa (single-celled creatures), and who emphasizes that "organisms must be studied in wild environments that challenge the organism to observe intelligent behaviour. It is perhaps no accident that the plant behaviour described in this article was largely published in ecological journals."[32]

* * *

And then there is *Cuscuta*, also called dodder or strangleweed. The plant has no roots, nor does it effectively photosynthesize. It gets its food by parasitizing other plants. On sprouting, the seedling has less than a week's worth of stored food. If it doesn't find a host within that time, it dies. But how does it find prey? By recognizing volatiles released by tomato and other species of host plants. Or, as those who are not human supremacist might put it, "smelling."

If you plant a seedling equidistant from a plant it prefers to eat, like a tomato, and a plant it doesn't, like wheat, the plant will grow toward the scent of the tomato. On reaching it, the dodder will encircle the prey and grow haustoria, roots that dig into the plant and begin to extract nutrients.[33]

* * *

Or there are tomatoes. Did you know that some plants, like tomatoes or potatoes (especially the wild variants) are carnivorous? No, I'm not talking about pitcher plants, bladderworts, sundews, and the other plants we know are carnivorous. It ends up that some plants, like tomatoes and potatoes, capture and kill small insects using sticky hairs along their stems. They do this as a way of feeding themselves. The insects decay, fall to the ground, and fertilize the soil at the plants' base.[34]

* * *

Of course, there are those who'd be quick to claim that none of this shows any plant intelligence at all. They'd argue that this is all mechanical, it's no different from the body producing white blood cells when confronted by an infection. I'd argue that you not consciously thinking about your white blood cells doesn't mean your body is also not thinking. The body is making those choices, and doing so with intelligence. But there's another point to be made here: when one plant communicates the presence of pests to another plant, how different is that from you communicating to another human that a swarm of mosquitoes is descending upon the both of you? Oh, but when plants do it, it's just distress pheromones being automatically released, kind of like a rush of adrenalin, except that everyone else can perceive it. When you do it, however, you're *choosing* to swat wildly at your face and arms.

There's yet another point, which is to ask, how would human supremacists mechanistically explain the ability of plants to make decisions based on future conditions?

I'm guessing they'd say the plants are detecting current factors in the environment that we simply cannot yet understand, but will be able to understand someday. These environmental factors cause hormonal changes that cause differential growth in plants. Actually I don't need to guess. The person who literally wrote the book on (and called) *Plant Physiology* has been described by Michael Pollan as "confident that eventually the plant behaviors we can't yet account for will be explained by the action of chemical or electrical pathways, without recourse to 'animism.'"[35] Problem solved. Supremacism maintained.

Whew!

Part of the problem is too often, too many people seem to believe that being able to track the biochemical or bioelectrical processes by which some decision is made or action taken implies that there is no volition in these choices or actions. But just because you understand that electrical signals traverse nerves, causing muscles to tighten or loosen as you move a pen, doesn't mean that writing *Hamlet* didn't require thought or creativity. Just because we understand that there is a chemical basis for brain function doesn't mean we don't think.

And just because you don't consciously tell yourself to sweat doesn't mean you can't think, "Wow, I'm really hot." And just because you don't consciously choose to shiver doesn't mean you can't say to your friends, "Damn, it's cold." Likewise, just because you can explain how a hard frost harms plants doesn't mean the plants don't talk about the weather.

A bigger part of the problem is that it's impossible to prove another being's subjectivity. You have only your own experience, and you cannot know for an indisputable fact that anyone else subjectively exists, that anyone else has experiences. For all you *know*, everyone but you could be projections of your own mind. When I was a teenager my friends and I used to play this game quite often. I'd say, "Ron, you are a figment of my imagination. I'm the only person who exists. Everything you do is because I imagine that's what you'll do." Since we were teenaged boys, he'd respond by socking me in the arm. I'd say, "I imagined you'd do that."

Unfortunately, we have an entire culture that belongs in the sad joke; is it solipsistic in here, or is it just me?

Or everyone else could be automatons programmed to act in certain ways, with programming sophisticated enough to cause them to sometimes "act" in ways that surprise you. When we figure out everything using our big and complex brain, there will be no more surprises!

This is not hyperbole. This is mechanistic science.

Here's another human supremacist response, and perhaps you can play spot the tautology on your own. We know that plants aren't intelligent because they don't have brains. Do you see it? Because we think humans think with our brains, we humans have decided that brains and central nervous systems are the only way that *anyone* can think and are therefore necessary for cognition. Therefore plants, not having brains and central nervous systems, cannot think. It's been proven. Thank goodness our supremacy withstood that one!

Think about it. What if I said to you that because quite often my sexual pleasure involves a penis ejaculating, all sexual pleasure experienced by anyone must involve a penis ejaculating? Ridiculous and self-centered, right? Sex has evolved on this planet in a myriad of beautiful and ecstatic ways, as different for flowering plants as it is for great apes as it is for shellfish as it is for fungi. And what if I said that just because my breathing involves lungs, all breathing must involve lungs (fish and trees and insects all say hello)? How do you know it isn't the same for intelligence?

It all comes down to this: the fundamental assumption of supremacists is that until proven otherwise—and, in fact, long after—supremacists presume the other is not a subjective being. It's what men and whites and the civilized have been doing since the beginnings of their respective supremacies. It's what supremacists do.

Why don't we flip that on its head? Why don't we assume that, until proven otherwise, others *are* subjective beings? Answer: because the truth is this: the only thing that matters is that we're number one. It makes it so much easier to rationalize exploiting everyone.

Years ago I asked a mechanistic scientist what would be sufficient proof to convince him that nonhumans can think, and that we can communicate with each other. He said, "If you asked the creature to do something against its [sic] nature, and it [sic] did so."

Can you see the problems with this?

What is its nature? Wouldn't this be the first round of the same sort of moving target as the brain-mass-to-body-ratio fiasco? You ask a dog to do something against her nature, and after a few training treats she'll roll over. The mechanist says, "Well, rolling over on command is obviously in her nature." So now you train her so when you point your finger at her and say "bang" she will stand on her hind legs, then fall over and lie still. The mechanist says, "But that's obviously in her nature, too, or she wouldn't do it." So you train her to keep a treat sitting on her nose until you say the phrase, "You can eat it," after which she tosses it in the air, then catches it. The mechanist says, "That's what I've been trying to say. Dogs are capable of being trained. That's not communication or sentience. That's just dogs responding mechanically to physical stimuli in a fashion that has been consciously molded by you. You're the only creative subjective force here. It's as though you fashioned a curvy clay track down which a marble could roll, and then you saying 'bang' is the equivalent of you starting the marble down the track. You've shown me no communication, no sentience on anyone's part but your own, as the teacher."

So what is or is not the creature's perceived "nature" will change according to what is required for the human supremacist to retain the self-perception of supremacy.

CHAPTER THREE

MOVING THE GOALPOSTS

There was no arguing his belief that what most men consider their rationally selected actions are in fact idiosyncratic responses (again, established during the decisive experiences of childhood) that have grown strong enough through repeated use, to overpower other urges and reactions—that have won, in other words, the mental battle for survival.

CALEB CARR

I just read an article in *Wired* titled "Fish Photographed Using Tools to Eat." It begins, "Professional diver Scott Gardner has captured what are believed to be the first images of a wild fish using a tool. The picture . . . shows a foot-long blackspot tuskfish smashing a clam on a rock until it cracks open, so the fish can gobble up the bivalve inside.

"Tool use was once thought to be exclusive to humans, and was considered a mark of our superior intelligent and bulging brains. In recent decades, though, more and more animals have shown an ability to work with tools and objects.

Elephants pick up branches with their trunk to swat flies and scratch themselves, a laboratory crow improvised a hooked tool from a wire to extract an insect and primates use sharpened sticks as spears, rocks to smash nuts and sticks to poke into ant nests.

"Tool use in fish, however, is much more rare, and there's never been any photo or video evidence to prove it—until now. 'The pictures provide fantastic proof of these intelligent fish at work using tools to access prey that they would otherwise miss out on,' said Culum Brown of Macquarie University in Sydney in a press release.

'It is apparent that this particular individual does this on a regular basis judging by the broken shells scattered around the anvil.'"

I bring this up not so much because it's surprising that some fish use tools—it isn't, particularly, at least to me—as to highlight the responses to this article at *Wired* online and to an excerpt of it at *Field & Stream*.[36] A fair number of the responses—far more than I would have guessed—accept this as tool use. One person wrote, "At first what the fish is doing doesn't seem that remarkable, but a little thought reveals its activity is actually highly complex. First of all, it's demonstrating a grasp the shell contains something worth the effort to acquire; then it's demonstrating a grasp of both the shell and the rock's differing properties, such that the properties of one can be used to counteract the properties of the other, (which mightn't sound that much until you realise it's demonstrating a grasp [that] seaweed, say, or sand can't provide the same function); then it's demonstrating a grasp [that] the two objects must be manoeuvred and made to act in relation to each other in a particular way, or the shell will remain intact; then it's demonstrating a grasp that initial failure can be overcome by tenacity and persistence; it's also demonstrating memory, either of a previous time it succeeded, or when it witnessed another fish succeeding; it's also demonstrating foresightedness, i.e., the ability to conceive and carry out a certain project for a certain reward." Someone else wrote, "Actually a photo was taken some years ago of salmon lining up along a suburban street during a rainstorm. They needed to get from water on one side to water on the other. They waited until a car passed, leaving a wake in the rain, then used the wake to wiggle across. They were lined up as if waiting for a bus." Another wrote, "I've seen blue jays dropping walnuts in front of cars to be run over. I've seen another bird picking up a snake many times and dropping it on the road in front of our car. [I've seen this as well.] Cars and roads . . . nut crackers and snake killers . . . I never thought of it." Yet another mentioned vultures who search for the right size rocks to throw at ostrich eggs in order to break them open.

And there was this: "Regardless of how smart the fish is, you can't ignore the important fact that it still doesn't have hands. This is about the highest possible level of fish tool use I can imagine a fish ever achieving (even if it was the smartest thing on earth), short of coercing another marine life form to do something for it." I mention this one because it makes clear something that will be a theme in this book, and something that is certainly a theme in this culture: the conflation of tool use, intelligence, or both with domination, or the ability to coerce another.

Another theme in this book, of course, is the moving target for what human supremacists believe constitutes intelligence. We've already seen this in the brain size discussion, and in the mechanistic scientist saying a nonhuman would have to act outside of its nature in order to convince him of its intelligence. Here's another example. In response to the fish using a tool, someone commented that "tool usage isn't a true sign of intelligence. Tool creation is. The ability to perceive of a future use for something and then creating a tool to match your needs is a sign of forethought, deep pondering and the mechanics of the universe, meaning u understand who u are, what u can do, where u are and where u might end up in the future, and in addition an ability to think abstractedly. . . . Best human example is the 'Rambo' style survival knife, multi functions built into the handle, ferocious-looking blade and oversize too. but also created in the last leg of the red danger-time and showing as much people's fears of external danger from other intelligent people as practicality." I have to admit I'd never before considered the Rambo style survival knife as the *ne plus ultra* of human creativity, but I think he has a point. And one final comment: "That doesn't seem much like a fish using a tool. It seems more like the fish was hungry and just busted open the clam. . . . I do not think that a fish could be dumb enough to eat your bait and get hooked yet is smart enough to use a tool."

I sometimes wonder something similar about members of this culture. I do not think a human could be stupid enough to swallow and get hooked on the belief that you can destroy a planet and live on it, yet be smart enough to use a tool.

❋ ❋ ❋

One reason fish strike at hooks is that they, like all of us, make decisions based on cost/benefit and risk/reward analyses. I'm hungry and need food. Some food chunks are fatal, but nearly all are nutritious. I need to weigh the risk of this particular piece of food killing me versus the gain of nutrition.

Humans make similar decisions every day. I'm hungry, so I'm going to drive to the grocery store. I'm going to get into a big metal missile and hurtle myself frighteningly close to other big metal missiles at high rates of speed. More than 30,000 people died in the United States last year in automobile accidents. I'm making a calculated decision, risking my life to get some food. And then there are the risks assumed once I've reached the grocery store, like buying potato chips and chicken nuggets.

We all make these sorts of decisions, so it's not a sign of stupidity on the part of fish when the decision ends up killing them.

That's one of the heartbreaking things about life: one decision can change everything. This was true when I was in the car wreck that broke my mom's neck; had we decided to stop for the evening five miles earlier, she would not now be functionally blind. Likewise, had a fly I earlier heard buzzing frantically as she was progressively bound by sticky silken threads decided a few moments earlier to go left instead of right, she would not have been caught by a spider. Had the snail not been exactly where she was and had I not unwittingly put my foot where I did . . .

Fish are really smart. They have good memories. Fish who have been caught are generally more difficult to catch again. And if fish are nearly caught by a predator at a specific place, they may avoid that place for several months. There are fish who can remember the human call announcing food for at least five years. We've been told that goldfish have a two-second memory. Not true. They can remember the color of a tube for dispensing food a year later. Other fish can remember signals associated with food for months. They can learn how to avoid traps, and if presented with the same trap eleven months later, still know how to avoid it.

Fish have complex social relationships. They remember the behavior of others in their groups and change their own behavior accordingly, for example avoiding fish who have bullied them. They also choose to associate with fish who are better rather than worse foraging partners.

Fish understand properties of transitivity. Scientists set up fights between males of a certain species of cichlid, and had other males watch. They learned that if the fish in question witnessed fish A beat up fish B who beat up fish C who beat up fish D who beat up fish E, they would consistently choose to associate with fish D over fish B, even though each had beaten up another once and been beaten up once. They choose to associate with the least dominant one.

Also, fish can deceive each other.

And they can learn from each other. Scientists captured some French grunts and released them in a new spot. Many grunts travel daily from sleeping to eating areas and back. The newly transplanted grunts followed the native grunts, and when the scientists removed the native grunts, the transplanted grunts continued to forage and rest at the places they'd been taught. Fish also learn from each other what are good food sources, and how to avoid predators. They can learn the scent of a predator by being exposed to that smell at the same time they see another fish who is frightened.

Of course fish can cooperate, swimming in shoals or schools, and hunting in packs. They sometimes and in many contexts work with fish of other species. For example, if a roving coralgrouper sees prey hiding somewhere the grouper can't get to, she might visit a local moray eel and shake her head outside the eel's lair. The eel knows this is an invitation to hunt. The grouper leads the eel to the hiding place, the eel heads on in, and either catches the prey or flushes it out for the grouper.

And we've all heard of cleaner wrasses and cleaner shrimp, right? Wrasses are tiny fish (and the shrimp are, well, shrimpy) who swim into the mouths of other fish. On purpose. Even those with big teeth. Why? To clean the big fish's teeth. Wrasses (and cleaner shrimp) eat the food that gets stuck there, and swim into the fish's gills and all over their bodies to eat parasites and clean off torn or worn scales. The big fish open their mouths wide, and float patiently (and presumably blissfully, like you would if you were getting a nice gentle exfoliation from your teeny tiny friend).

Fish, predator and prey alike, line up outside the wrasse cleaning station, like humans waiting in queue for buses, or to use better examples, waiting at the dental hygienist's office, or for a pedicure. The wrasse will make its way down the line.

If the wrasse sees a non-local fish passing by, the wrasse will put this one at the front of the line, presumably knowing that if she doesn't take care of this one right now, the fish will pass on to another cleaning station, as opposed to the locals, who may perceive this station as their only option. Frankly that seems ungrateful to me, and not a business model I would choose, but wrasse have been doing this for millions of years, and I've never once stuck my head in a moray eel's gaping and toothy maw, so in this case I'll refrain from giving advice, something that is very difficult for a middle-aged white male to do. I hope I don't pull a muscle. Sometimes wrasse get cheeky, and take a tiny bite out of the fish they're cleaning (their clients? Patients? Customers? Neighbors?). For obvious reasons they're far more likely to do this to herbivores than carnivores. And if the client takes umbrage and starts to swim away, the wrasse will swim after the client and make a big fuss, then give the client a little back massage with her fins as a way of apology.

❀ ❀ ❀

The more I learn about the real world, the more wonderful I think it is, and the more honored I am to be here.

❀ ❀ ❀

Oh, and by the way, some spiders use stones as tools as well.[37]

❀ ❀ ❀

Remember the guy above, who thought the best human example of tool-making is the *Rambo*-style knife, and who said, in true moving target fashion, that tool-making and not tool use is the "true" sign of intelligence? Well, I wonder what he'd say about cockatoos making tools.

Researchers put a cockatoo outside a wire cage that contained a nut. They gave the cockatoo a thin piece of wood that was too wide to fit into the cage. He pretty quickly figured out how to break off a sliver slender enough to slide in. The cockatoo put one end in his mouth and used the other to bat the nut out of the cage. The cockatoos who watched him learned from him how to do this, and when given the opportunity to do it, not only used his technique, but refined it by figuring out better ways to make the sliver from the original thin slab.

A few of the comments on the article were decent. But a lot were precisely what we'd expect. One commenter, who had the screen name John Gault, insisted that those who care about nonhuman welfare need to "get a life," and suggested that the fact that humans can reflect on whether nonhumans are intelligent is a sign that humans have superior, to use his spelling, "intelect" (to which someone reasonably responded, "How are you so sure that other species don't reflect on things?" and to which I might add, "And how are you so sure they don't ask themselves whether or not humans are intelligent?") He also commented—and this might be my favorite human supremacist comment of all time—"Most men display sentienticity way above any animal."

❋ ❋ ❋

Don't ever let anyone tell you that most men don't display more sentienticity than every other being on the planet.

❋ ❋ ❋

Did you know that caterpillars self-medicate? No, not for depression (that we know of), although sometimes being a caterpillar can be tough: not only do human supremacists systematically poison you and destroy your habitat, but lots of species of flies and wasps lay eggs inside of caterpillars. The eggs hatch, then feed on the caterpillar's internal organs before bursting forth like tiny versions of the creature from *that scene* in *Alien*—you know the one. That's enough to make a caterpillar depressed right there. But instead of wallowing in depression, caterpillars do something about it: they eat leaves from senecio plants and others, and flood their bodies with alkaloids, chemical compounds that are often toxic (and sometimes have pharmacological uses; caffeine, morphine, and cocaine are all alkaloids, as are nicotine, atropine, muscimol, quinine, psilocybin, ergotamine, yohimbine, and strychnine). Scientists don't know whether the alkaloids kill the parasites directly, or if they boost the caterpillars' immune systems. The point is: it works. And it ends up that infected caterpillars eat more alkaloid-containing leaves than do those who don't need the cure. Healthy caterpillars still eat small doses of the leaves, presumably to make themselves taste bad. The caterpillars have to know how much to eat. If they eat too much, they end up shortening their own lives, probably much like many humans who over-medicate.

I think this is all pretty cool—and also not particularly surprising. I told one friend who replied, "I thought we knew that more or less all creatures make decisions about what to eat based on the needs of their bodies." But never fear: the journalists and scientists want to make sure we know that even though caterpillars are making choices, no, they certainly couldn't be thinking. *National Geographic News* reports, "The new finding challenges the idea that self-medication is restricted to relatively intelligent creatures that [sic] are capable of learning, such as primates [later we will explore the reality that basically everyone is capable of learning, with the possible exception of human supremacists]. For example, chimps can learn which drugs to take to cure their ills and can pass on that knowledge to others. Something much simpler is probably happening in insects. When a woolly bear is infected by parasites, its immune system may react by altering taste receptors so that the animals crave more alkaloids, Elizabeth Bernays of the University of Arizona said. Insects 'have a system that's based on changes to their taste system, rather than the cognitive ability of their brains,' she added."[38]

But even if the caterpillars are eating more of these plants because the alkaloids taste good to them, that still doesn't mean the caterpillars aren't making decisions. They have to find the plants, they have to figure out how to get to the plants, and they have to go.

❋ ❋ ❋

This is a great example of what I mentioned about how too many people seem to believe that being able to track the biochemical or bioelectrical processes by which some decision is made or action taken implies there's no volition in these choices or actions.

I remember in high school I sometimes went to a restaurant that offered all-you-can-eat steaks. I'm sure they lost money on every teenaged boy who walked through the door, and even more on me and my remarkably inefficient metabolism. But I soon became convinced they had an angle. Since the first and second steaks always tasted better than the third and fourth steaks, and the fifth and sixth steaks tasted pretty awful, I figured the cooks paid attention to how many times you returned to the grill, and the more times you returned the worse steaks they gave you. This would discourage people from eating a lot of steaks and costing them a lot of money.

I've never said I'm always the sharpest claw in the paw.

In my defense, it only took me three or four visits to realize that the cooks were not in fact counting my steaks, but that food tastes better when we're hungry.

Taste exists for a reason. Food not tasting good is one of the ways we know we're full, and don't need to eat more. Likewise, some berries taste good because they provide a lot of energy. We co-evolved with plants and others who grow substances that when eaten and metabolized provide us with different amounts or kinds of nutrition, so we evolved such that we experience a pleasant sensation when we ingest, for example, huckleberries or blackberries, as a way of encouraging us to do so. Likewise, many substances that aren't good for us taste bad. There's a reason huckleberries taste better to most of us than shit. I've always presumed, however, that shit tastes good to dung beetles, some species of flies, some species of worms, and so on. Likewise, the wood I've tasted in my life is pretty bland—and I just learned that a lot of fast and/or processed foods contain some form of powdered cellulose, which in plain terms is wood pulp, so there's a chance you've eaten some yourself—but I'm guessing a lot of termites would have a different opinion. So the termites might have preferred the table to even that delicious first steak. And maybe they like their wood pulp with special sauce, lettuce, cheese, and so on at McDonald's.

My point is that the fact that the perception of taste in caterpillars might change when they require a specific food so that baby wasps or flies—who presumably prefer the taste of living caterpillar guts to wood pulp, or they'd burrow their way right out of the caterpillar and into a nearby Burger King—don't eat them alive, in no way implies that the caterpillars aren't making decisions. It isn't so different from a decision you or I might make a decision not to have that fourth piece of cheesecake, since it might not taste *quite* as good as the first one.

❊ ❊ ❊

Why is it so difficult for us to accept that others besides ourselves have rich interior lives?

I guess because that would mean our moral universe just got a whole lot more complex.

* * *

A female botanist friend just gave me another example of a human supremacist moving target. A while back she briefly dated a vegan. The relationship came to a sudden halt when he insisted she convert. After she refused, he accused her of human supremacism, saying, "You hold yourself above animals." Then he asked, "How can you eat someone who is sentient?"

She responded, "Everyone is sentient."

"Plants aren't sentient."

She was way ahead of me on the plant sentience thing. She laid into him with all the stuff I just mentioned, and plenty more.

When she finished, he shook his head and said, "Even more important than sentience is suffering. How you could eat someone who suffers?"

"What makes you think plants don't suffer?"

"No central nervous system."

We've been through that one before. She used her equivalent to my "since my sexual pleasure sometimes includes ejaculation, then everyone's must" comment (a precursor to my comment, really; I learned more of my plant sentience chops from her than the other way around). Then she asked, "So plants don't feel distress? They don't have physiological responses to discomfort or danger?" She summarized for him some of the literature on the personal and communal effects of various forms of stress on plants.

He said, "Those are nothing but physiological responses."

Even though they were in a café, she let out a scream. He leapt from his chair. Everyone stared at her. She said, "What you just did was nothing but a physiological response."

"What's your point?"

"You jump up and open your eyes wide when startled. Plants have different responses to sudden threat. Yet because their response is different than yours, they must not get scared, right? Likewise, you suffer when you're thirsty, and plants suffer when they're thirsty. Yet because plants' manifestations of this suffering don't follow the same patterns as do yours, they must not suffer, right? Likewise, they suffer when predators come after them, and they fight back. But since they don't suffer like you suffer, and they don't fight back like you'd fight back, they must not really suffer, and they must not fight back. They communicate with their neighbors. But they don't communicate the way you communicate with your neighbors, which means they don't communicate at all. When they're pulled from the ground, they suffer, but because they don't suffer the way you suffer, they must not suffer at all."

He said, "No central nervous system equals no pain, no suffering, no sentience." Then he smiled smugly and said, "No brain, no pain."

The last thing she ever said to him was, "We have much deeper differences than my belief in plant sentience."

❀ ❀ ❀

She later told me she thought part of the hesitation of so many people to acknowledge that everyone else is alive and everyone else is sentient is that they are fearful of living in a world that is nearly infinitely complex, and nearly infinitely morally complex. It's much more convenient to live in a world where your morality is based on a clearly defined hierarchy, with you at the top. To interact with a machine is less complex and less morally complex than to interact with a community.

She's right. She's also right when she points out what a wonderful experience it is to find yourself in a world of infinite complexity, and infinite moral complexity. It's as though before, you could only move in one or two dimensions, and now suddenly you're living and moving in three dimensions, then four, then five, and on and on, and you then wonder how you never went mad from the claustrophobia of earlier having restricted yourself to only one or two dimensions.

❀ ❀ ❀

To not be able to conceptualize or accept the existence of any form of intelligence that does not resemble your own does not seem to me very intelligent. And to not be able to conceptualize or accept the existence of any form of suffering that does not resemble your own does not seem to me very compassionate or empathetic. In both cases it seems quite the opposite: not only unintelligent, and not only showing a complete lack of anything remotely resembling imagination, but cruel.

※ ※ ※

Almost twenty years ago I interviewed Cleve Backster about plant intelli-gence. No, he wasn't a botanist. He was one of the world's experts on the use of polygraphs, or lie detectors. I know that sounds like an odd con-nection, but listen to his story, and the connection will become clear. Just after World War II he was a CIA interrogation specialist, and founded The Agency's polygraph school. In 1960 he left the CIA and formed the Backster School of Lie Detection, to instruct police officers. This school is the longest running polygraph school in existence.

Backster could name the moment the focus of his life changed for-ever, from lie detection to plant intelligence: early in the morning on February 2, 1966, at thirteen minutes, fifty-five seconds of chart time for a polygraph he was administering. He had threatened the subject's well-being in hopes of triggering a response. The subject had responded electrochemically to this threat. The subject was a plant.

Here's his story: "I wasn't particularly into plants, but there was a going-out-of-business sale at a florist on the ground floor of the building, and the secretary bought a couple of plants for the office: a rubber plant, and this dracaena cane. I had done a saturation watering—putting them under the faucet until water ran out the bottom of the pots—and was curious to see how long it would take the moisture to get to the top. I was especially interested in the dracaena, because the water had to climb a long trunk, and then to the end of long leaves. I thought if I put the galvanic-skin-response detector of the polygraph at the end of a leaf, a drop in resistance would be recorded on the paper as the moisture arrived between the electrodes. . . . I noticed something on the chart resembling a human response on a polygraph: not at all what I would have expected from water entering a leaf. Lie detectors work on the principle that when people perceive a threat to their well-being, they physiologically respond in predictable ways. If you were conducting a polygraph as part of a murder investigation, you might ask a suspect, 'Was it you who fired the shot fatal to so and so? If the true answer were *yes*, the suspect will fear getting caught lying, and electrodes on his or her skin will pick up the physiological response to that fear. So I began to think of ways to threaten the well-being of the plant. First I tried dipping a neighboring leaf in a cup of warm coffee. The plant, if anything, showed what I now recognize as boredom—the line on the chart just kept trending downward.

"Then at thirteen minutes, fifty-five seconds chart time, the imagery entered my mind of burning the leaf. I didn't verbalize; I didn't touch the plant; I didn't touch the equipment. Yet the plant went wild. The pen jumped right off the top of the chart. The only new thing the plant could have reacted to was the mental image.

"I went into the next office to get matches from my secretary's desk, and lighting one, made a few feeble passes at a neighboring leaf. I realized, though, that I was already seeing such an extreme reaction that any increase wouldn't be noticeable. So I tried a different approach: I removed the threat by returning the matches to the secretary's desk. The plant calmed right back down.

"Immediately I understood something important was going on. I could think of no conventional scientific explanation. There was no one else in the lab suite, and I wasn't doing anything that might have provided a mechanistic trigger. From that split second my consciousness hasn't been the same. My whole life has been devoted to looking into this."

He called what the plant was doing "primary perception." He found that not only plants were capable of this: "I've been amazed at the perception capability right down to the bacterial level. One sample of yogurt, for example, will pick up when another is being fed. Sort of like, 'That one's getting food. Where's mine?' That happens with a fair degree of repeatability. Or if you take two samples of yogurt, hook one up to electrodes, and drop antibiotics in the other, the electroded yogurt shows a huge response at the other's death. And they needn't even be the same kind of bacteria. The first Siamese cat I ever had would only eat chicken. I'd keep a cooked bird in the lab refrigerator and pull off a piece each day to feed the cat. By the time I'd get to the end, the carcass would be pretty old, and bacteria would have started to grow. One day I had some yogurt hooked up, and as I got the chicken out of the refrigerator to begin pulling off strips of meat, the yogurt responded. Next, I put the chicken under a heat lamp to bring it to room temperature, and heat hitting the bacteria created more huge reactions in the yogurt."

I asked how he knew he wasn't influencing it.

"I was unaware of the reaction at the time. I had pip switches all over the lab, and whenever I performed an action, I hit a switch, which placed a mark on a remote chart. Only later did I compare the reaction of the yogurt to what had been happening in the lab."

"Did the yogurt respond again when the cat started to eat?"

"Interestingly enough, bacteria appear to have a defense mechanism such that extreme danger causes them to go into a state similar to shock. In effect, they pass out. Many plants do this as well. If you hassle them enough they flatline. The bacteria apparently did this, because as soon as they hit the cat's digestive system, the signal went out. There was a flatline from then on."

Cleve continued, "I was on an airplane once, and had with me a little battery-powered galvanic response meter. Just as the attendants started serving lunch, I pulled out the meter and said to the guy next to me, 'You want to see something interesting?' I put a piece of lettuce between the electrodes, and when people started to eat their salads we got some reactivity, which stopped as the leaves went into shock. 'Wait until they pick up the trays,' I said, 'and see what happens.' When attendants removed our meals, the lettuce got back its reactivity. I had the aisle seat, and I can still remember him strapped in next to the window, no way to escape this mad scientist attaching an electronic gadget to lettuce leaves.

"The point is that the lettuce was going into a protective state so it wouldn't suffer. When the danger left, the reactivity came back. This ceasing of electrical energy at the cellular level ties in, I believe, to the state of shock that people, too, enter in extreme trauma."

"Plants, bacteria, lettuce leaves . . ."

"Eggs. I had a Doberman Pinscher back in New York whom I used to feed an egg a day. One day I had a plant hooked up to a large galvanic response meter, and as I cracked the egg, the meter went crazy. That started hundreds of hours of monitoring eggs. Fertilized or unfertilized, it doesn't matter; it's still a living cell, and plants perceive when that continuity is broken. Eggs, too, have the same defense mechanism. If you threaten them, their tracing goes flat. If you wait about twenty minutes, they come back.

"After working with plants, bacteria, and eggs, I started to wonder how animals would react. But I couldn't get a cat or dog to sit still long enough to do meaningful monitoring. So I thought I'd try human sperm cells, which are capable of staying alive outside the body for long periods of time, and are certainly easy enough to obtain. I got a sample from a donor, and put it in a test tube with electrodes, then separated the donor from the sperm by several rooms. The donor inhaled amyl nitrate, which dilates blood vessels and is conventionally used to stop a stroke. Just crushing the amyl nitrate caused a big reaction in the sperm, and when the donor inhaled, the sperm went wild.

"So here I am, seeing single-cell organisms on a human level—sperm—that are responding to the donor's sensations, even when they are no longer in the same room as the donor. There was no way, though, that I could continue that research. It would have been scientifically proper, but politically stupid. The dedicated skeptics would undoubtedly have ridiculed me, asking where my masturbatorium was, and so on.

"Then I met a dental researcher who had perfected a method of gathering white cells from the mouth. This was politically feasible, easy to do, and required no medical supervision. I started doing split-screen videotaping of experiments, with the chart readout superimposed at the bottom of the screen showing the donors activities. We took the white cell samples, then sent the people home to watch a preselected television program likely to elicit an emotional response—for example, showing a veteran of Pearl Harbor a documentary on Japanese air attacks. We found that cells outside the body still react to the emotions you feel, even though you may be miles away.

"The greatest distance we've tested has been about three hundred miles. Astronaut Brian O'Leary, who wrote *Exploring Inner and Outer Space*, left his white cells here in San Diego, then flew home to Phoenix. On the way, he kept track of events that aggravated him, carefully logging the time of each. The correlation remained, even over that distance."

"The implications of all this . . ."

He interrupted, laughing. He said, "Yes, are staggering. I have file drawers full of high quality anecdotal data showing time and again how bacteria, plants, and so on are all fantastically in tune with each other. And human cells, too, have this primary perception capability, but somehow its gotten lost at the conscious level."

"How has the scientific community received your work?"

"With the exception of scientists at the margins, like Rupert Sheldrake, it was met first with derision, then hostility, and mostly now with silence. At first they called primary perception 'the Backster Effect,' perhaps hoping they could trivialize the observations by naming them after this wild man who claimed to see things missed by mainstream science. The name stuck, but because primary perception can't be readily dismissed, it is no longer a term of contempt.

"What's the primary criticism by mainstream scientists?"

"The big problem—and this is a problem as far as consciousness research in general is concerned—is repeatability. The events I've observed have all been spontaneous. They have to be. If you plan them out in advance, you've already changed them. It all boils down to this: repeatability and spontaneity do not go together, and as long as members of the scientific community overemphasize repeatability in scientific methodology, they're not going to get very far in consciousness research.

"Not only is spontaneity important, but so is intent. You can't pretend. If you say you are going to burn a plant, but don't mean it, nothing will happen. I hear constantly from people in different parts of the country, wanting to know how to cause plant reactions. I tell them, 'Don't do anything special. Go about your work; keep notes so later you can tell what you were doing at specific times, and then compare them to your chart recording. But don't plan anything, or the experiment won't work.' People who do this often get equivalent responses to mine, and often win first prize in science fairs. But when they get to Biology 101, they're told that what they have experienced is not important.

"There have been a few attempts by scientists to replicate my experiments . . . but these have all been methodologically inadequate. . . . It is so very easy to fail. . . . And let's be honest: some of the scientists were relieved when they failed, because success would have gone against the body of scientific knowledge."

I said, "For scientists to give up predictability means they have to give up control, which means they have to give up Western culture, which means it's not going to happen until civilization collapses under the weight of its own ecological excesses."

He nodded, then said, "I have given up trying to fight other scientists on this, because I know that even if the experiment fails they still see things that change their consciousness. People who would not have said anything twenty years ago often say to me, 'I think I can safely tell you now how you really changed my life with what you were doing back in the early seventies.' These scientists didn't feel they had the luxury back then to rock the boat; their credibility, and thus their grant requests, would have been affected."

I asked if there were alternative explanations for the polygraph readings. I'd read that one person suggested his machine must have had a loose wire.

He responded, "In thirty-one years of research I've found all my loose wires. No, I can't see any mechanistic solution. Some parapsychologists believe I've mastered the art of psychokinesis—that I move the pen with my mind—which would be a pretty good trick itself. But they overlook the fact that I've automated and randomized many of the experiments to where I'm not even aware of what's going on until later, when I study the resulting charts and videotapes. The conventional explanations have worn pretty thin. One such explanation, proposed in *Harper's*, was static electricity: if you scuffle across the room and touch the plant, you get a response. But of course I seldom touch the plant during periods of observation, and in any case the response would be totally different."

"So, what is the signal picked up by the plant?

"I don't know. I don't believe the signal, whatever it is, dissipates over distance, which is what we'd get if we were dealing with electromagnetic phenomenon. I used to hook up a plant, then take a walk with a randomized timer in my pocket. When the timer went off, I'd return home. The plant always responded the moment I turned around, no matter the distance. And the signal from Phoenix was just as strong as if Brian O'Leary were in the next room. Also, we've attempted to screen the signal using lead-lined containers, and other materials, but we can't screen it out. This makes me think the signal doesn't actually go from here to there, but instead manifests itself in different places. All this, of course, lands us firmly in the territory of the metaphysical, the spiritual."

I said, "Primary perception suggests a radical redefinition of consciousness."

"You mean it would do away with the notion of consciousness as something on which humans have a monopoly?" He hesitated a moment, then continued, "Western science exaggerates the role of the brain in consciousness. Whole books have been written on the consciousness of the atom. Consciousness might exist on an entirely different level."

I asked whether he had worked with materials that would normally be considered inanimate.

"I've shredded some things and suspended them in agar. I get electric signals, but not necessarily relating to anything going on in the environment. It's too crude an electroding pattern for me to decipher. But I do suspect that consciousness goes much, much further. In 1987 I participated in a University of Missouri program that included a talk by Dr. Sidney Fox, then connected with the Institute for Molecular and Cellular Evolution at the University of Miami. Fox had recorded electric signals from protein-like material that showed properties strikingly similar to those of living cells. The simplicity of the material he used and the self organizing capability it displayed suggest to me that bio-communication was present at the earliest states in the evolution of life on this planet. Of course the Gaia hypothesis—the idea that the earth is a great big working organism, with a lot of corrections built in—fits in nicely with this. I don't think it would be a stretch to take the hypothesis further and presume that the planet itself is intelligent."

I asked how his work has been received in other parts of the world.

"The Russians and other eastern Europeans have always been very interested. And whenever I encounter Indian scientists—Buddhist or Hindu—and we talk about what I do, instead of giving me a bunch of grief they say, 'What took you so long?' My work dovetails very well with many of the concepts embraced by Hinduism and Buddhism."

"What is taking us so long?"

"The fear is that, if what I am observing is accurate, many of the theories on which we've built our lives need complete reworking. I've known biologists to say, 'If Backster is right, we're in trouble.' It takes a certain kind of character and personality to even attempt such a questioning of fundamental assumptions. The Western scientific community, and actually all of us, are in a difficult spot, because in order to maintain our current mode of being, we must ignore a tremendous amount of information. And more information is being gathered all the time. For instance, have you heard of Rupert Sheldrake's work with dogs? He puts a time-recording camera on both the dog at home and the human companion at work. He has discovered that even if people come home from work at a different time each day, at the moment the person leaves work, the dog at home heads for the door.

"Even mainstream scientists are stumbling all over this bio-communication phenomenon. It seems impossible, given the sophistication of modern instrumentation, for us to keep missing this fundamental attunement of living things. Only for so long are we going to be able to pretend it's the result of 'loose wires.' We cannot forever deny that which is so clearly there."

❀ ❀ ❀

Faced with what Backster was saying, I had several options. I could believe he's either a crackpot or lying, as is everyone else who has ever made similar observations. I could believe that what he was saying is true, which would validate many things I had experienced but would require that the whole notion of repeatability in the scientific method be reworked, along with preconceived notions of consciousness, communication, perception, and so on. Or I could believe that he had overlooked some strictly mechanistic explanation. But seriously, static electricity, humidity, a loose wire? Are those the best excuses the human supremacists can come up with?

Or, and here's the real solution, I could see for myself.

❀ ❀ ❀

Backster hooked up a plant. We chatted. I watched the paper roll out of the recorder. I couldn't correlate the movement of the pen with anything I was feeling, or with the conversation. A cat started to play with the plant. The oscillations of the pen seemed to increase in magnitude, but I couldn't be sure. Halfheartedly, I suggested burning the plant. No response from the plant. Cleve responded, "I don't think you really want to, and besides, I wouldn't let you."

We moved to another part of the lab, and he put yogurt into a sterilized test tube, then inserted a pair of sterilized gold electrodes. We began again to talk. The pen wriggled up and down, and once again seemed to lurch just as I took in my breath to disagree with something he said.

But I couldn't be sure. When we see something, how do we know if it is real, or if we are seeing it only because we wish so much to believe? The same is true, of course, for not seeing events.

Cleve left to take care of business elsewhere in the building. The line manifesting the electrical response of the yogurt immediately went flat. I tried to fabricate anger, thinking of clearcuts and the politicians who legislate them, thinking about abused children and their abusers. Still flat. Either fabricated emotions don't count (as Cleve had suggested), or it's a sham, or something else was terribly wrong. Perhaps the yogurt wasn't interested in me.

Losing interest myself, I began to wander the lab. My eyes fell on a calendar, which on closer inspection I saw was actually an advertisement for a shipping company. I felt a surge of anger at the ubiquity of advertising. Then I realized—a spontaneous emotion! I dashed to the chart, and saw a sudden spike corresponding to the moment I'd felt the anger. Then more flatline.

And more flatline. And more. Again I began to wander the lab, and again I saw something that triggered an emotion. This was a poster showing a map of the human genome. I thought of the Human Genome Diversity Project, a monumental study hated by many Indigenous people and their allies for its genocidal implications (Backster is not affiliated with or particularly a fan of the program; I later found he simply likes the poster). Another surge of anger, another dash to the chart, and another spike in the graph, from instants before I started to move.

❀ ❀ ❀

If your experience of the world is at variance with what this culture inculcates you into believing should be your experience of the world, what do you do?

❀ ❀ ❀

Many people respond by denying their own experience.

Of course. That's the point of a supremacist philosophy.

❊ ❊ ❊

I just read a blog account of someone who was "suffering from a serious slug problem" in her kitchen. One night she accidentally stepped on one, getting its guts all over her bare foot. So, "traumatised and utterly disgusted," she "went on a revenge-driven, murderous killing-spree with the sodium chloride."

Please note that in addition to the redundancy—killing sprees by definition involve murder—*she* is the one who is traumatized—never mind the slug she crushed—and *she* is the one who must seek revenge on the others of the species for one member having the temerity to happen to put its body in the path of her foot. This is a window into the hatred of nature that accompanies human supremacism. It is analogous to those ranchers who kill every wolf they see because one wolf ate one calf the ranchers were running in the wolf's home. Or more accurately, it's analogous to a rancher killing every wolf he sees, then tracking down and killing the rest of a pack when a wolf bleeds on him.

Back to the woman with slugs in her kitchen. Watching "the slugs writhe around for several minutes following administration of the salt treatment" further traumatized her, because "it looked painful; death by dehydration seems like a pretty unpleasant way to die."

How to deal with the trauma? Supremacism, once again, to the rescue: "In an attempt to alleviate my contrition I tried to tell myself that it killed them reasonably quickly and they didn't suffer for long. Then I began to wonder . . . do they actually suffer at all? Do slugs and other such gastropod molluscs actually have a nervous system that is sufficiently developed to generate the sensation of pain as we know it?"

Although the author acknowledges that "higher [sic] invertebrates—some worms, flies and our friend *Limax*—have quite highly developed nervous systems, believe it or not—only a few notches down the evolutionary ladder from ourselves. [Please note the Great Chain of Being reference, in full human supremacist glory.] They have highly developed sensory organs which send nerve impulses along sensory neurones to clusters of neurones in the head. These are called central ganglia, and are essentially a very primitive brain. Information is then relayed to muscles in different parts of the body through a nerve cord (not dissimilar to the vertebrate spinal cord) that runs from head to tail of the animal, and allows changes in behaviour. So, actually, the nervous system organisation in these invertebrates is rather similar to our own. Not great for the ego, eh?"

It's not great for the ego if you base your self-worth on having nothing in common—not even the rudiments of your nervous system—with other residents of this planet.

The author went searching for experiments where scientists had tortured mollusks, ostensibly to determine whether the nonhumans felt pain. She discovered that "the first thing to happen when you roast a snail is that it retracts into its shell to minimise immediate damage. Secondly, if the snail remains on the hot-plate for more than 30 seconds or so, it will protract from its shell, secrete a thick, insulating, yellow mucous, and display searching movements—very sensible—in an attempt to get to somewhere that is not as hot. These searching movements involve contraction of the foot (the part in contact with the hot-plate), and repeated turning of the body from side to side." She acknowledges that this fits with what humans would do in a similar situation, and fits as well with what she had observed in the slugs: "The first thing that the slugs do is contract their bodies to about half of their normal length, and curl up at the edges. Then, they begin their characteristic writhing around that I described; moving their gait rapidly from one side to the other in an attempt to find somewhere less salty. Death comes too swiftly to allow the secretion of a mucous, but I bet that if you were to put the slugs on a non-lethal salty surface, there would be a mucous secretion, just like in snails."

Her conclusion comes in standard nonsensical human supremacist fashion: "My own feeling is that slugs DON'T feel pain in the sense that we know it, and my reasons for thinking this are thus. The sensation of 'pain' is not generated directly at the area of damage. In vertebrates such as ourselves, damage stimulates pain receptors in tissues, and electrical impulses are sent to the brain. The brain then integrates and interprets the information, and makes you *feel* pain in the area that you've damaged. But there's the key point—pain is a *feeling* that is generated by the *brain*: specifically, if you're interested, in two regions known as the peri-aqueductal grey matter and the nucleus raphe magnus."

I'm so glad we have science to tell us not to believe the writing that is happening before our eyes.

The author even acknowledges that morphine has an effect on the pain response of other creatures such as lobsters, but then lets us know that this doesn't mean that lobsters feel pain. It just means that morphine affects their responses to pain (which, according to her, they don't feel).

Don't bother trying to figure out the logic. It doesn't really hang together. It doesn't have to. Neither logic nor evidence were ever going to be allowed to lead where they may, but rather were going to be tortured into shape to serve her supremacism. Near the end the author reveals the real point of the whole damn article: "At least, I can sleep safe in the knowledge that I did not cause the slug the most unbearable agony that I initially thought I had."[39]

❄ ❄ ❄

This, succinctly stated, is the central point and most important function of any supremacist philosophy.

CHAPTER FOUR

COMPLEXITY AND ITS OPPOSITE

I believe nature is intelligent. The fact that we lack the language skills to communicate with nature does not impugn the concept that nature is intelligent. It speaks to our inadequacy for communication.

PAUL STAMETS

Science deals with but a partial aspect of reality, and . . . there is no faintest reason for supposing that everything science ignores is less real than what it accepts. . . . Why is it that science forms a closed system? Why is it that the elements of reality it ignores never come in to disturb it? The reason is that all the terms of physics are defined in terms of one another. The abstractions with which physics begins are all it ever has to do with.

J.W.N. SULLIVAN

Physical science will not stop short of a reduction of the universe and all it contains to the basis of mechanics; in more concrete terms, to the working of a machine.

CARL SNYDER

Throughout this discussion I can't stop thinking about one of the most important passages I've ever read. In Neil Evernden's life-changing book *The Natural Alien*, he describes how some vivisectionists "adopted a routine precaution: at the outset of an experiment they would sever the vocal cords of the animal on the table, so that it could not bark or cry out during the operation. This is a significant action, for in doing it the physiologist was doing two other things: he was denying his humanity, and he was affirming it. He was denying it in that he was able to cut the vocal cords and then pretend the animal could feel no pain, that it was merely the machine Descartes claimed it to be. But he was also affirming his humanity in that, had he not cut the cords, the desperate cries of the animal would have told him what he already knew, that it *was* a sentient, feeling being, and not a machine at all.

"That act is an appropriate metaphor for the creation of a biological scientist out of a nature-lover. The rite of passage into the scientific way of being centres on the ability to apply the knife to the vocal cords, not just of the dog on the table, but of life itself. Inwardly, he must be able to sever the cords of his own consciousness. Outwardly, the effect must be the destruction of the larynx of the biosphere, an action essential to the transformation of the world into a material object subservient to the laws of classical physics. In effect, he must deny life in order to study it."[40]

❊ ❊ ❊

In the late 1970s and early 1980s a few scientists discovered what trees have known for a very long time, that plants communicate. In one study, a scientist from the University of Washington fed leaves from Sitka willows being eaten by tent caterpillars and webworms to captive insects, and learned that the insects grew more slowly than normal. The willows were altering the composition of their leaves to stunt the growth of the predators. Next, he found that leaves from other willows nearby—those not themselves being eaten—also caused the insects to grow more slowly. The nearby plants were changing their leaf composition as well. Around that same time, a couple of scientists from Dartmouth discovered that poplar and sugar maple seedlings were also capable of similar communication.

The response by the scientific community to even this slight threat to human supremacism—the threat being that trees may share a trait with us[41]—followed the pattern that believers in supremacisms often follow when their supremacisms are threatened. Out came the tautologies and poor thinking, the clutching at straws, the bullying. Plants don't have nervous systems, and therefore they must not be able to do those things that in animals require nerves (never mind that there could be, and evidently are, other means by which others achieve these ends). Just as Cleve Backster must have had wires (or screws) loose, or must have shuffled his feet across the carpet then zapped the plant with static electricity, the researchers from Dartmouth must have designed their studies poorly, and the researcher from the University of Washington must have in some unspecified way accidentally spread some unspecified disease to the captive insects. The UW researcher couldn't get funding to replicate his study—that's certainly one way to guarantee a lack of repeatability—and eventually left science altogether to run a bed and breakfast.[42]

Let's jump forward to 2013, when at least a few scientists are learning something else that plants have known more or less forever, that plants of different species communicate. If you harm sagebrush, for example, it gives off signals to which tobacco plants respond. Harming cucumbers causes responses by chili peppers and lima beans. As one journalist says, "It turns out almost every green plant that's been studied releases its own cocktail of volatile chemicals, and many species register and respond to these plumes." This same writer calls these chemical communications "a universal language." And it's not only other plants who respond. That journalist continues, "Plants can communicate with insects as well, sending airborne messages that act as distress signals to predatory insects that [who] kill herbivores. Maize attacked [sic] by beet armyworms releases a cloud of volatile chemicals that attracts wasps to lay eggs in the caterpillars' bodies. The emerging picture is that plant-eating bugs, and the insects that [who] feed on them, live in a world we can barely imagine, perfumed by clouds of chemicals rich in information. Ants, microbes, moths, even hummingbirds and tortoises . . . all detect and react to these blasts."[43]

And did I mention that plants communicate not only through these volatile chemicals, but also with "electrical pulses and a system of voltage-based signaling that is eerily reminiscent of the animal nervous system"?[44]

The response by human supremacists continues to be much the same as it ever was. One supremacist calls plant intelligence "a foolish distraction," while another says discussions of it are "the last serious confrontation between the scientific community and the nuthouse on these issues."

A third says that those who discuss plant intelligence are suffering from "over-interpretation of data, teleology, anthropomorphizing, philosophizing, and wild speculations."[45]

When a supremacism of any sort is one of the unquestioned beliefs acting as a real authority of that culture, defenders of that supremacism nearly always perceive any questioning of any part of that supremacism as a "foolish distraction." They generally portray themselves—and quite often perceive themselves—as defenders of reasonableness and sanity, and perceive those questioning their supremacism as having come from "the nuthouse." This is as true of human supremacists today as it was of defenders of race-based chattel slavery and as it was of defenders of the witch trials. And because the beliefs that underlie their supremacisms are unquestioned, proponents of supremacisms can say without intentional irony that they're not philosophizing or participating in wild speculations. Because their supremacist perspective is unquestioned—and the supremacists would prefer it remain that way—all questioning of that supremacism by definition will be classed as speculation, and all speculation on that subject will be discouraged. Of course, speculating about ways to escalate the ability of one's superior class to exploit all inferior classes is seen as innovation, creativity, and a sign of one's intelligence and superiority. So, discuss your perception of nonhuman sentience, and you're a foolish distraction from the nuthouse who is speculating; figure out a way to use cyanide to extract gold from rocks and leave behind a poisoned landscape, and you're a fucking hero and a shining example of human ingenuity.

Teleology is one of those philosophical buzzwords that mechanistic scientists often throw out to try to nerd-bully into silence those with whom they disagree. It's analogous to a Christian telling you that you've just said something blasphemous: in each case, the real message is that you're expressing an opinion that violates dogma.

In the case of plant communication, and of this modest attempt to help people remember that this world really does have a voice, or rather uncounted millions of voices, there's a sense in which the scientist's use of the word is an attempted slur, a sense in which it's an attempt to limit discourse, a sense in which it's unintentionally ironic, and a sense in which it's indicative of a destructive mindset.

It's obvious that *teleology* was thrown out as a slur, and not to promote discourse, because it's irrelevant to the specific discussion of whether plants do or don't communicate. One dictionary defines teleology as "the explanation of phenomena by the purpose they serve rather than by postulated causes."[46] Another states, "A teleological school of thought is one that holds all things to be designed for or directed toward a final result, that there is an inherent purpose or final cause for all that exists. It is traditionally contrasted with metaphysical naturalism [sic], which views nature as lacking design or purpose. In the first case form is defined by function, in the second function is defined by form. Teleology would say that a person has eyes because [s]he has the need of eyesight (form follows function), while naturalism [sic] would argue that a person has sight simply because [s]he has eyes, or that function follows form (eyesight follows from having eyes)."[47] Some definitions suggest that a teleological perspective implies the existence of a God or gods, or some sort of design to nature. Most definitions contrast human actions, which may under this rubric have purpose, from nonhuman actions, which under this rubric may not. Here's a not atypical example: "Within material reality, only human artifacts possess intelligent form and intelligent functionality or purpose. Measurable biological patterns lack intelligibility in themselves. Similarly, biological functionality is not truly functionality, but merely resembles the functionality of human engineering."[48]

Really? That's a lot of narcissistic assumptions and self-glorifying tautologies packed into less than forty words.

The point here is that in a discussion centered on whether or not plants communicate, it's not really important whether plants needed to communicate, and then evolved ways of doing so (form followed function); or whether plants gave off scents, and the plants who perceived these scents had a greater chance of survival (function followed form). In both explanations the plants communicate. Any discussions of final causes or intelligent designers are just as irrelevant to the question of *whether* plants communicate as they are to whether humans communicate. We don't need final causes. We don't need a designer. To teleologize or not to teleologize, that is not the question.

His use of the word *teleology* is limiting not only for the obvious reason that it was an attempt to discourage research into plant communication—and maybe if he was lucky the researchers would quit science altogether and start running a bed and breakfast—but also because even if we take the word *teleology* to not be a slur, but at face value, and merely presume the scientist meant that teleology, philosophizing, and speculation are, without value-judgment, to be deemed no part of science, then that by definition still limits exploration. Scientific philosopher Richard Dawkins would probably be surprised to learn that philosophizing should never be associated with science. And just because the scientific philosophy and speculations espoused by this human supremacist scientist claim to disallow teleology—just because an anti-teleological science has as one of its central tenets that the world has no intelligent form or functionality or purpose, and that only humans are able to create intelligently—doesn't mean that there are no other ways to know or understand anything. And we will never know those ways so long as we're not allowed to explore them.

The scientific philosopher Francis Bacon—and I guess we can presume that this means he philosophized, and presumably speculated, so I guess it's only those researching plant communication who aren't supposed to do those things—spoke of putting nature on the rack and torturing her to extract her secrets. Evidently it is only nature we are supposed to examine closely, not the assumptions of mechanistic science.

But that shouldn't really come as a surprise.

Bacon was very clear about why he wanted to torture nature. He wrote, "My only earthly wish is . . . to stretch the deplorably narrow limits of man's dominion over the universe to their promised bounds." Human supremacists want this same thing today. He also wrote, "I am come in very truth leading you to Nature with all her children to bind her to your service and make her your slave. The mechanical inventions of recent years do not merely exert a gentle guidance over Nature's courses, they have the power to conquer and subdue her, to shake her to her foundations."[49]

And this is exactly what science and technology—all guided by human supremacism—have done.

Given what Bacon wrote, it shouldn't surprise us that scientists have argued so strongly against nonhuman sentience; the last thing any slave-master wants is to consider the possibility that his slaves have lives of their own, and do not wish to be bent to his will.

At least Bacon, whom I hate more than almost any other Western philosopher, had enough integrity—and it's certainly not much—to explicitly acknowledge he wanted to torture nature. Nowadays they're not always so direct. As they in all truth shake nature to her foundations, scientists call it hydraulic fracking or geophysical exploration, or, to speak of another foundation being shaken, genetic modification.

The scientist's use of the word *teleology* is also unintentionally ironic, for a couple of reasons. The first is that if we use as our definition of teleology "the explanation of phenomena by the purpose they serve rather than by postulated causes," then a lot of science is pretty damn teleological. Physicists don't know precisely what *causes* gravity, or for that matter light or electricity. Instead they've developed equations that describe how these phenomena function. And it's not just physics. Medical researchers don't actually know how the majority of medicines work. In many cases they don't understand the causal connections, but merely understand that when they give medicine A to patient B, symptoms C and D subside, and in six percent of cases the patient gets side effects E through J. So they understand the drugs in terms of their purpose. And then there are geology and paleontology, where scientists as a matter of course take end points and then work backwards to postulate what might have caused this rock formation or that fossil. I know that the use of the word *purpose* is still strictly forbidden, since one of the commandments of human supremacist science is that in all the universe, only humans and their projects are allowed to be described as having purpose, but my point is that it seems that when we aren't talking about nonhuman sentience, it's perfectly acceptable in science to work backwards from known effects to "speculate" as to causes, so long as these causes continue to support the notion that humans are the only ones with agency, so long as the causes continue to support the Great Chain of Being.

When scientists use the word *teleology* as a slur, they show that they've forgotten the difference between an assumption and a fact. It is taken as a given by much of the scientific community—and indeed, much of this human supremacist culture—that nonhumans can't communicate, can't think, don't appreciate life, and so on. But this is merely a given, and is not in any non-tautological way shown. This is all ironic, given science's self-proclaimed status as a bastion of open and free inquiry.

And as I said at the beginning of this book, this assumption of human supremacy is killing the planet. I'm certainly not the first to write of how the so-called Enlightenment and its applied twin the industrial revolution have combined to become a possibly fatal disaster for life on earth. The former has destroyed our perception of the world as full of life and meaning and purpose and wild sentiences; the latter has used the tools of the former to accomplish this destruction in the real, physical world. The former provides the philosophical foundation and methods for the latter.

Recall what that journalist from *Nature* [sic] online said to me: "If nature were to cease to exist, nature itself would not notice, as it is not conscious (at least in the case of most animals and plants, with the possible exception of the great apes and cetaceans) and, other than through life's drive for homeostasis, is indifferent to its own existence. Nature thus only achieves worth through our consciously valuing it." He has obliterated his perception of the natural world and replaced it with his own dead projections. He has cut the vocal cords of his own empathy and, in his own perception, of the world. And how much easier is it to destroy some other we do not perceive as having inherent meaning, some *thing* that does not (in this supremacist's perception) even care if *it* lives or dies?

Further, if you accept teleology as a slur, then you presumably accept the dreadful notion that "only human artifacts possess intelligent form and intelligent functionality or purpose. Measurable biological patterns lack intelligibility in themselves. Similarly, biological functionality is not truly functionality, but merely resembles the functionality of human engineering." And if you accept that "biological functionality is not truly functionality," then you can come to disbelieve that, for example, salmon have irreplaceable and true functionality regarding forests, or that rivers have irreplaceable and true functionality regarding salmon, and so forth. And if you come to disbelieve in these biological functionalities, it means, well, for one thing it means you're insane, since you're not believing in physical reality, and for another, you may come to believe that you can kill off the salmon without harming the forest, or that you can murder a river without harming the salmon. You may come to believe that as the only one who is able to create true functionality, you can destroy, as modern humans are doing, the "biological functionality" of the oceans to metabolize carbon dioxide into oxygen, and a) survive; and b) replace this functionality by one of your own creation, which would, of course, be the only true functionality. You may come to believe that forests can't manage themselves, but that you can manage forests. You may come to believe that after you destroy glaciers, you can create your own and replace their evidently untrue functionality with a true functionality of your own.[50] You may come to believe that the world cannot survive without your interference, while the truth is that the world cannot survive your arrogant interference.

There is not one natural community on the planet that has been managed by human supremacists which that management has not either destroyed or is in the process of destroying.

Human supremacists posit humans as the smartest beings around (in fact, really, the only smart ones). Members of this culture contrast themselves positively with members of other cultures, who are more "primitive," less sophisticated, or let's cut to the chase, less intelligent than they are. And scientists are often portrayed as the brainiest of the brainiacs in this culture, the smartest of the smart, the most discerning of the discerning, the sapiens of the sapiens. But I think it's pretty fucking stupid to assume you're the only one who can think, and it's even more stupid to forget that your assumption is nothing but an assumption. And it's even stupider still to continue to think you're smarter than anyone else as your culture destroys life on this planet, fueled in great measure by your perception of yourself as the most intelligent and meaningful being in the universe. Actually the only intelligent and meaningful being in the universe.

The scientist also used the word *anthropomorphizing*. The word means "to ascribe human characteristics to an animal, inanimate object, force of nature, etc." So many of the insane unquestioned presumptions of human supremacism lie in one not-so-little word. Depending on the context, it presumes a human ownership of intelligence, the ability to communicate, the ability to suffer, the ability to feel emotions, and so on. Any perception of these among nonhumans is to be eradicated. Never mind that these are all natural attributes. The primary justification for this separation is the Great Chain of Being. Which is no reasonable justification at all.

The use of the word *anthropomorphizing* as a slur by human supremacists and especially mechanistic scientists is even more ironic than their complaints about *teleology*, in that the human supremacists often use machine language when discussing nonhumans. The Enlightenment and the Scientific Revolution are *based* on seeing the world as a giant machine, and nonhumans as "beast machines," as Descartes put it. Modern scientific discourse is, as would be expected considering the basis of modern science, based on machine language. But machines are a uniquely human-made project. You can't get more anthropomorphic than to describe the world in mechanistic terms, to project a human construct onto the real, living world. This is one reason I do not speak, for example, of ecosystems, but rather natural communities.

Their use of the word *anthropomorphizing* is even more ironic and absurd than I've so far made it seem: isn't it just a tad anthropomorphic to require that everyone else's intelligence, response to pain, sorrow, joy, and so on, resemble one's own?

Or maybe it's just narcissistic.

Michael Pollan asked plant neurobiologist Stefano Mancuso "why he thinks people have an easier time granting intelligence to computers than to plants. ([Prominent botanist] Fred Sack told me [Pollan] that he can abide the term 'artificial intelligence,' because the intelligence in this case is modified by the word 'artificial,' but not 'plant intelligence.' He offered no argument, except to say, 'I'm in the majority in saying it's a little weird.') Mancuso thinks we're willing to accept artificial intelligence because computers are our creations, and so reflect our own intelligence back at us. They are also our dependents, unlike plants: 'If we were to vanish tomorrow, the plants would be fine, but if the plants vanished . . .' Our dependence on plants breeds a contempt for them, Mancuso believes. In his somewhat topsy-turvy [sic] view, plants 'remind us of our weakness.'"[51]

Mancuso's point is a good one, to which we will return.

The evolutionary ecologist Monica Gagliano wanted to determine whether plants are capable of learning. Given that industrial humans have destroyed or are destroying every natural community they try to manage (read, steal from and try to control), yet they still continue to try to manage (steal from and try to control) every natural community they can find instead of leaving them alone; and given that industrial civilization is killing the planet, and yet most industrial humans don't seem interested in even *acknowledging* that industrial civilization is killing the planet, much less getting rid of it, and thereby allowing life on this planet to continue, I'd be more interested in determining whether industrial humans are capable of learning. Be that as it may, she came up with a fascinating way to conduct her experiment.

Pollan writes, "She focused on an elementary type of learning called 'habituation,' in which an experimental subject is taught to ignore an irrelevant stimulus. 'Habituation enables an organism to focus on the important information, while filtering out the rubbish,' Gagliano explained to the audience of plant scientists. How long does it take the animal to recognize that a stimulus is 'rubbish,' and then how long will it remember what it has learned? Gagliano's experimental question was bracing: Could the same thing be done with a plant?

"*Mimosa pudica*, also called the 'sensitive plant,' is that rare plant species with a behavior [and I love the fact that he used the word *behavior*] so speedy and visible that animals can observe it; the Venus flytrap is another. When the fernlike leaves of the mimosa are touched, they instantly fold up, presumably to frighten insects. The mimosa also collapses its leaves when the plant is dropped or jostled. Gagliano potted fifty-six mimosa plants and rigged a system to drop them from a height of fifteen centimetres every five seconds. Each 'training session' involved sixty drops. She reported that some of the mimosas started to reopen their leaves after just four, five, or six drops, as if they had concluded that the stimulus could be safely ignored. 'By the end, they were completely open,' Gagliano said to the audience. 'They couldn't care less anymore.'

"Was it just fatigue? Apparently not: when the plants were shaken, they again closed up. '"Oh, this is something new,"' Gagliano said, imagining these events from the plants' point of view. 'You see, you want to be attuned to something new coming in. Then we went back to the drops, and they didn't respond.' Gagliano reported that she retested her plants after a week and found that they continued to disregard the drop stimulus, indicating that they 'remembered' [and I see no need for scare quotes] what they had learned. Even after twenty-eight days, the lesson had not been forgotten. She reminded her colleagues that, in similar experiments with bees, the insects forgot what they had learned after just forty-eight hours. Gagliano concluded by suggesting that 'brains and neurons are a sophisticated solution but not a necessary requirement for learning,' and that there is 'some unifying mechanism across living systems that can process information and learn.'"

As Cleve Backster said, "It seems impossible, given the sophistication of modern instrumentation, for us to keep missing this fundamental attunement of living things. Only for so long are we going to be able to pretend it's the result of 'loose wires.' We cannot forever deny that which is so clearly there."

He underestimated the power of denial. By now we can predict the response of the supremacists. One scientist's reasoned response was "Bullshit." And the tautologies. Oh, the tautologies. Only animals can learn because, well, only animals can learn. Plants can "evolve adaptations" but never learn. Never mind that we don't normally talk about beings evolving adaptations within a single generation, unless you want to say that children, uh, evolve an adaptation into reading when parents read to them, in which case we're back to learning, only using fancier words. Further, as Gagliano said, "How can they be adapted to something they have never experienced in their real world?" She noted that some plants learned faster than others, evidence that "this is not an innate or programmed response." Another scientist said that there's nothing to discuss, because no matter what happened, "it's not learning." And why is it not? Evidently because, well, it just isn't. So there.

The relevant question is whether these scientists are capable of learning. Perhaps we can devise an experiment where we drop them from a height of fifteen centimeters every five seconds until they change their behavior.

Pollan continues, "Someone objected that dropping a plant was not a relevant trigger, since that doesn't happen in nature. Gagliano pointed out that electric shock, an equally artificial trigger, is often used in animal-learning experiments. Another scientist suggested that perhaps her plants were not habituated, just tuckered out. She argued that twenty-eight days would be plenty of time to rebuild their energy reserves."

Gagliano has been trying to get the article published, but so far ten journals have rejected it. Pollan quotes her saying, "'None of the reviewers had problems with the data.' Instead, they balked at the language she used to describe the data. But she didn't want to change it. 'Unless we use the same language to describe the same behavior'—exhibited by plants and animals—'we can't compare it,' she said."[52]

This makes me happy.

CHAPTER FIVE

VALUE-FREE SCIENCE

Capitalism as we know it couldn't exist without science. And science as we know it has been formed and deformed by capitalism at every step of the way.

STANLEY ARONOWITZ

If feminist psychology is correct, the very concept of scientific "objectivity" as a disciplined withdrawal of sympathy by the knower from the known, is a male separation anxiety writ large. Written, in fact, upon the entire universe.

THEODORE ROSZAK

I'll tell you two things about much of this plant research, however, that break my heart. The first is that even a few of the researchers themselves—and I'm certainly not talking about Mancuso or Gagliano—believe that the plant communication they're studying could not actually be plant communication. Oh, sure, they understand that plants disperse and receive and respond to various chemicals—and Mancuso, who's trying to make a dictionary of these chemicals, estimates a vocabulary of about three thousand terms,[53] which frankly compares favorably with some humans I've known—but then insist for ideological reasons that this communication could not be any sort of communication. As science journalist Kat McGowan writes, "For both [plant researchers] Karban and Heil, the outstanding question is evolutionary: Why should one plant waste energy clueing in its competitors about a danger? They argue that plant communication is a misnomer; it really might just be plant eavesdropping. Rather than using the vascular system to send messages across meters-long distances, maybe plants release volatile chemicals as a faster, smarter way to communicate with themselves—Heil calls it a soliloquy. Other plants can then monitor these puffs of airborne data."[54]

Wait! What just happened? Remember that curse word of scientists: *anthropomorphization*? Now let's get this straight: some middle-aged white males believe that when plants speak, they're mainly doing this to hear themselves talk, and if anyone else happens to derive benefit from their ramblings, that's all just coincidental? Middle-aged white males saying this? Project much?

I'm making a joke, kind of, but the fact remains that these scientists are projecting their worldview onto the plants. They asked, "Why should one plant waste energy clueing in its competitors about a danger?" This question manifests pretty much everything that's wrong with this culture, wrong with science, wrong with this culture's relationship with the natural world, and wrong with relationships between humans in this culture. This question succinctly shows why and how this culture is killing the planet.

I'll answer the question. Why should one plant "waste" energy clueing in its "competitors" about a danger? Because the plants are smart enough to understand that they're not the only creatures on the planet, and that their very survival requires the well-being of all these others. Plants in a redwood forest, for example, understand that a redwood forest consists of more than just redwoods, that it consists also of alders and cedars and firs and ferns and fungi and bears and otters and salmon and caddis flies. Industrial humans, including these researchers, don't seem to understand that. And industrial humans, including especially foresters, don't seem to understand that a tree farm of Douglas firs is not a forest. These plants understand that life is not a game of *Risk*, and they understand that the point of life is not for one group to eliminate every other group and conquer the world. They know that to do so would be immoral, insane, suicidal, and stupid. They know enough not to measure "superiority" by the ability to destroy all "competitors" but rather by the ability to improve the capacity of the landbase to support them. They understand something understood by Indigenous humans but understood by almost none of the civilized, that, as the Dakota writer Vine Doloria wrote, "Life is not a predatory jungle, 'red in tooth and claw,' as Western ideology likes to pretend, but a symphony of mutual respect in which each player has a specific part to play. We must be in our proper place and play our role at the proper moment."

My niece is visiting. I read her the scientists' question, "Why should one plant waste energy clueing in its competitors about a danger?"

She threw her hands into the air and exclaimed, "These people! Don't they understand the importance of community?"

Evidently not.

I shared the quote (and the rest of the analysis) with another friend who responded, "This says everything you need to know about their worldview, and nothing about the real world. Why don't they use the word *neighbor* instead of *competitor*?"

Probably because they're members of this exploitative culture. It shouldn't surprise us that members of the same culture that gave us capitalism as the dominant economic model—based as it is on the insane notion that selfish individuals all attempting to maximally exploit each other will somehow create stable and healthy human communities (never mind that it never has and functionally cannot)—would give us variants of the selfish gene theory as the dominant biological model—based as *it* is on the equally insane notion that selfish individuals all attempting to maximally exploit each other will somehow create stable and healthy natural communities (never mind that it never has and functionally cannot). Both are justifications for what the dominant culture does: steal from everyone else. Absent is the reality of how communities survive and thrive. These must be absent, if members of our culture are going to feel good about themselves as they steal from and destroy everyone else, and as they ultimately kill the planet.

It's actually worse than this: not only must the reality of how communities survive and thrive be absent, acknowledgment that communities even *exist* must be fundamentally absent. In the 1980s, neoliberal icon Margaret Thatcher said, "There is no such thing as society. There are individual men and women, and there are families." How different is this, really, than scientists believing that the most important unit of evolution is the individual (or even more perversely, the individual gene that happens to be carried by this individual, that happens to be within some larger collection of individuals, whom this individual is driven to exploit by selfish genes), and not communities. Communities don't exist, except as collections of individuals from whom (or rather which) we must steal. It's capitalism projected onto the natural world.

But wait, I can hear the scientists say, this is *not* capitalism projected onto the natural world, because we're just describing how the world really works. We're not philosophizing or speculating or projecting or anthropomorphizing. This is reality!

Just yesterday I heard a scientist say on television, "Science *is* truth."

And of course that's one of the problems with science. It allows exploiters to pretend they're describing reality when they're speculating and projecting with the worst of them. And part of the point of any exploitative philosophy is to make the exploitation seem natural or inevitable. Thus it is pleasing for kings and their allies to propagate the notion that kings are placed on thrones by a God who looks quite like them. It is pleasing for men and their allies to propagate the notion that they are placed on their smaller more familial thrones by a God who also looks quite like them. It is pleasing for those who wish to steal land from American Indians to propagate the notion that it is their Manifest Destiny to overspread the continent, and it is pleasing as well for them to believe their way of life is superior to all of these others. It is pleasing for those who wish to exploit others to create a Great Chain of Being and to place themselves at its earthly top. It is pleasing for those who wish to "exploit natural resources" to create a philosophy, a worldview, an ideology, and a theology which declares the world to consist of "natural resources," not other beings, and to deride evidence to the contrary as "speculation" or "philosophizing" or "anthropomorphizing." It is pleasing for those who perceive themselves as superior to all others to create various and mutable rationales for this superiority, whether it is to generate a mythology where you're created in the image of an omnipotent God or to create a mythology where your notion of what is true is based on your ability to enslave others, as Richard Dawkins puts it when he says that "science bases its claims to truth on its spectacular ability to make matter and energy jump through hoops on command," in other words, when he makes clear that the very epistemology of this culture is based on the ability to enslave. It is pleasing to those who wish to exploit others to declare, as writer Charles Mann does, about a world "run by human beings for human purposes," that "anything goes. . . . Native Americans managed the continent as they saw fit. Modern nations must do the same." But there is a world of difference between indigenous peoples forming long-term relationships with their landbases, and ExxonMobil drilling for gas."[55]

The creation of these ideologies of domination not only eases or erases the consciences of the perpetrators, but makes resistance to these perpetrators seem futile. In the case of Christianity, for example, it's hard enough to fight the king and all who believe in his divine right of kingship without adding in the possibility that you're going against the Big Man Himself. And so far as science, I've often commented that science is a far better means of social control than Christianity, in that if you question Christianity you're merely consigned to a hell you don't believe in, but if you question science you must be just plain stupid.

So here's the point. It's extraordinarily useful for those whose lifestyles are based on the systematic exploitation of others to pretend that this exploitation is natural. Thus they needn't worry their consciences about this exploitation, which they no longer perceive as exploitation, and no longer perceive even as "just the way things are," but rather as completely expected. Inevitable. Natural.

And when your way of life is predicated on narcissism and community-destroying sociopathy—to the point of perceiving yourself the only one who really matters on the planet, killing the planet, and then using this planetary murder to validate your own self-perceived superiority—it's extraordinarily useful to pretend that evolution itself is driven by supremely narcissistic individual actions, and that community not only is not central but plays no effective role in evolution, and that any evidence to the contrary must be either ignored or derided as "speculation," "philosophizing," or, with an entirely-to-be-expected narcissism, "anthrompomorphization." Thus it becomes easy to pretend plant communication is soliloquy, and thus it becomes easy to destroy forests, grasslands, wetlands, rivers, oceans, the world.

It has long been clear to me that the most important elements in evolution—which really means in life—are biotic communities, who are themselves living beings. Just as your own body is made up of other living beings, some of whom share your DNA and the vast majority of whom do not, so, too, the larger living bodies of forests and grasslands and ponds and streams and rivers and oceans are made up of other smaller living beings, living beings whose lives are as precious to them as the larger being's is to it, and as yours is to you. And these smaller beings affect the health of the larger being. And just as your body is permeable, so, too are the bodies of these others. A river flows into a forest; water enters the body of the forest. The river flows out; water leaves the body of the forest. Both the river and the forest are alive. And when salmon spawn and die in this river, this is the river, these salmon, feeding the forest. And when trees drop their leaves into the water, this is the forest, these trees, feeding the river. When the river floods, the river and the forest feed each other.

What can seem destructive may not be. Bears girdle trees, which kills them. But forests need standing dead trees as homes for some of those who live there.[56] And dead trees can continue to feed everyone else, by slowly becoming soil as they are eaten by animals, fungi, other plants, bacteria, and so on, and in other ways as well.

Have you ever heard of what are called "Mother Trees"? These are big old trees who are connected to swaths of forest through the mycelial networks, and who help to feed and maintain the other trees—especially the younger ones—in that part of the forest. Even after the trunks die, the mother trees continue to feed these others as long as they can. As one enthusiastic description has it: "Counter to Darwin's 'survival of the fittest' theory,[57] Mother Trees do not compete for resources; rather, their presence ensures the healthy survival and diversity of younger, newer trees [and other plants], as they actively transfer vital nutrients and forest wisdom via an overlapping, interconnected, fungi-rich web of shared roots. If a Mother Tree is to die, she will consciously transfer her resources to her interlinked community of living trees before she fully collapses, knowingly 'passing her wand' to the next generation."[58]

If all the talk of "wisdom" and "consciously" and "passing her wand" freaks you out, we can instead speak the language of forestry: "Forest ecologist Suzanne Simard and her colleagues at the University of British Columbia have made a major discovery: trees and plants really do communicate and interact with each other. They discovered an underground web of fungi connecting the trees and plants of an ecosystem. This symbiotic web enables the purposeful sharing of resources, that consequently helps the whole system of trees and plants to flourish. 'The big trees were subsidizing the young ones through the fungal networks,' Simard says. 'Without this helping hand, most of the seedlings wouldn't make it.' Dr. Simard was led to the discovery by the observation of webs of bright white and yellow fungal threads in the forest floor. Many of these fungi were mycorrhizal, meaning they have a beneficial, symbiotic relationship with a host plant, in this case tree roots. Microscopic experimentation revealed that the fungi actually move carbon, water and nutrients between trees, depending upon their needs. At the hub of a forest's mycorrhizal network stand the 'Mother Trees'—large, older trees that rise above the forest, a concept illustrated in the movie *Avatar*. These 'Mother Trees' are connected to all the other trees in the forest by this network of fungal threads, and may manage the resources of the whole plant community. Simard's latest research reveals that when a Mother Tree is cut down, the survival rate of the younger members of the forest is substantially diminished."[59]

Forests, and the trees who live in them and are parts of them, know how absurd it is to ask, "Why should one plant waste energy clueing in its competitors about a danger?"

Who are the intelligent ones?

❊ ❊ ❊

One of the most elegant arguments I've seen against ruthless competition as the central driving force of evolution came, oddly enough, in the book *The Selfish Gene*, Richard Dawkins's hymn to ruthless competition; it's a projection onto the natural world of the same mindset that gave us neoliberal capitalism. At one point in the book he proposes a thought experiment in which a population of creatures coexists with ticks. These creatures cannot groom themselves, and if they don't have the ticks removed through groomings, the ticks can kill them. He writes, "Let the population consist of individuals who adopt one of two strategies. As in Maynard Smith's analyses, we are not talking about conscious strategies [of course, since mechanistic science is all about projecting a lack of consciousness onto the entire universe], but about unconscious behaviour programs laid down by genes [of course, since so much of mechanistic science is about naturalizing oppressive and exploitative behavior, in this case by blaming genes for selfish, community-destroying behavior]. Call the two strategies Sucker and Cheat.[60] Suckers groom anybody who needs it, indiscriminately. Cheats accept altruism from suckers, but they never groom anybody else, not even somebody who has previously groomed them. As in the case of the hawks and doves, we arbitrarily assign pay-off points. It does not matter what the exact values are, so long as the benefit of being groomed exceeds the cost of grooming. If the incidence of parasites is high, any individual sucker in a population of suckers can reckon on being groomed about as often as he grooms. The average pay-off for a sucker among suckers is therefore positive. They all do quite nicely in fact, and the word *sucker* seems inappropriate. But now suppose a cheat arises in the population. Being the only cheat, he can count on being groomed by everybody else, but he pays nothing in return. His average pay-off is better than the average for a sucker. Cheat genes will therefore start to spread through the population. [Please note that he presumes the cheater does so not because of personality, deformation of personality through trauma, or deformation of personality through narcissistic philosophy, but rather because genes told him to; the net effect of his language is to naturalize exploitation.] Sucker genes will soon be driven to extinction. This is because, no matter what the ratio in the population, cheats will always do better than suckers. For instance, consider the case when the population consists of 50 per cent suckers and 50 per cent cheats. The average pay-off for both suckers and cheats will be less than that for any individual in a population of 100 per cent suckers. But still, cheats will be doing better than suckers because they are getting all the benefits—such as they are—and paying nothing back. When the proportion of cheats reaches 90 per cent, the average pay-off for all individuals will be very low: many of both types may by now be dying of the infection carried by the ticks. But still the cheats will be doing better than the suckers. Even if the whole population declines toward extinction, there will never be any time when suckers do better than cheats. Therefore, as long as we consider only these two strategies, nothing can stop the extinction of the suckers and, very probably, the extinction of the whole population too."[61]

Dawkins then goes on to describe a third strategy he calls "grudgers," who will groom others when they first meet, but without reciprocity will never groom that individual again. This strategy ultimately wins out in his model, with "suckers" being eliminated and "cheats" being reduced to a small percentage.

But for me the real point had already been made, in his story of how "cheats" destroy previously stable communities of "suckers." I first read (and hated) *The Selfish Gene* in 1990. As I read that passage, in a park on a warm late-summer day in Spokane, Washington, I loudly exclaimed, "That's it exactly. Doesn't everyone else see it?" The other people in the park evidently did not, since they merely looked at me like I was a crazy man.

Despite the fact that Dawkins is with his work arguing for a selfish gene-induced innate sociopathy and a lack of communal responsibility, it seemed perfectly clear to me that Dawkins was here, combined with current events, making an elegant and concise argument in favor of cooperation as a primary mover of evolution, and the community as evolution's primary unit.

Do you see it? Do you see how his example of suckers and cheats makes the opposite point—as powerfully as is possible—to what he intended?

Let's try this then. Instead of one species with two strategies, let's pretend we have a hundred different populations of different species within some natural community. All of these species are, to for a moment use Dawkins's term, suckers, who groom others. But here this grooming can take the form of many actions besides pulling off ticks. So why don't we just call them "givers"? Perhaps this giving comes in the form of a fish, who, having ingested a certain species of parasite, follows the parasite's instructions to swim to the surface of the water and flash its belly to the sky. This makes it easier to catch, and a seabird ingests the fish as well as the parasite. The bird gives by pooping out the parasite's children, who are then eaten by someone else who is eaten by a fish, who then swims to the surface and flashes its belly. And the parasite gives by allowing the birds to eat: without them it would be too hard to catch fish and the birds would die, and the entire community would begin to unravel. Or let's take salmon. They give their bodies to a forest. The forest gives wood and soil to the river. The river gives soil and food to the ocean. The ocean gives water to the air (and food in the form of anadromous fish to forests). The air gives water to the forest. Everyone gives. The Mother Trees give. The voles who eat mushrooms in the forest give the spawn in their poop to the soil. The spawn in the soil joins with root tips of Douglas firs. The firs and the fungi feed each other, and together grow a tree who is a home for the voles, and the owls who eat the voles. Bears girdle the tree and kill it. Now it becomes homes for others. All give. From each according to its gifts, and the needs of the community. To each according to its needs and the needs of the community. These gifts can include their lives. For many, especially for the very young of some species, their lives are their only gift, as among tadpoles or many others, the overwhelming majority of them give gifts of their lives in the form of food not long after they are born.

Can you imagine a model of ecological sustainability like this? And can you see where I'm going with it? Even Richard Dawkins states in *The Selfish Gene* that a community of givers (or to use his term "suckers") would be ecologically stable, so long as it encountered no "cheats," whose presence would destabilize and then destroy the formerly stable community. In fact, a community of all "suckers" would be the richest and most fecund, as all would receive the most benefit.

Humans are in this gift economy, too. They give just as everyone else does. From each according to its gifts, and the needs of the community. To each according to its needs, and the needs of the community. Humans are completely integrated into the community. This was, for example, how the Tolowa lived where I live now. When humans are integrated they all, as Dawkins says in his example, "do quite nicely." And indeed, the word "sucker" does seem inappropriate.

Now, what does Dawkins's model state happens when cheats move into a previously stable community where givers live, to a forest, to a bay, to a grassland? Because these cheat are "getting all the benefits" and "paying nothing back"—and does this sound like the behavior of anyone we know?—they will deplete the "suckers" (the givers) until there is nothing left.

As we see.

The cheats will prosper at the expense of the givers, and eventually the cheats will so destroy the givers that they will destroy their own ability to cheat, and thereby wipe out themselves as well.

As we see.

Here's the thing: Dawkins has perfectly described what this culture of cheats is doing to the planet. If a primary argument for selfishness is that a world filled with givers would collapse when a cheat arrived, and the dominant culture is clearly a cheat who has arrived and is causing the world to collapse, wouldn't that in fact be an argument that prior to the arrival of the cheat the world just might have been full of givers?

And how did he think there got to be so many salmon in the first place, so many fecund forests, rich grasslands, vibrant marshes, rivers and oceans full of fish? Where did these come from? They came from the members of these forests, grasslands, marshes, rivers, oceans living and dying and making their homes better places by their lives and deaths. By all this giving.

And now the world is doing what his model predicts. What else does he think is happening, as salmon populations collapse, as do those of migratory songbirds, and as the oceans die?

Why doesn't everybody see this? I guess the answer might be that any culture that would kill the planet would use any means necessary, including, of course, philosophy, to avoid perceiving the consequences of its actions, and to ignore even the most straightforward logic.

As we see.

❊ ❊ ❊

Please note in addition to all of this that even according to the model used by Dawkins, the "cheats" need not be more intelligent nor in any other way superior to the "suckers" in order to effectively drive both of them to extinction. The "cheats" do so merely by cheating.

I'll be explicit: the fact that members of this culture have through cheating gained a competitive advantage over other humans and non-human beings in no way implies any form of greater intelligence or any other form of superiority. It implies what it is: cheating gains a competitive advantage at the cost of future extinction of those from whom the cheater is taking, and then also the cheater himself.

❊ ❊ ❊

How you perceive the world affects how you behave in and toward the world. If you perceive competition as the world's guiding principle, compete you will; if you perceive the world as being full of ruthless competitors you must overcome and exploit, you will do your part to ruthlessly overcome and exploit them. If, on the other hand, you perceive the world's guiding principle to be that of giving to the larger biotic community, you will give to the larger biotic community; if you perceive the world as being full of others who give to make it stronger, healthier, more alive, then you will do your part to make it so.

❀ ❀ ❀

I just read an article about how chimps consistently outperform humans in certain sorts of games that require they pay close attention to, and recognize patterns in, what the other player is doing.

The game the article describes is basically a matching game, where each player secretly chooses left or right on a computer screen the other can't see, and if the players match, then player A wins, and if they don't, then player B wins. Chimps imprisoned in laboratories outperformed sixteen university students from Japan and twelve men from Guinea. The chimps moved more quickly in each case to optimal strategies than did the humans. There was no difference in ineptitude between the humans from Japan and Guinea.

What possible implications might one draw from this?

Well, the first implication I might draw could be that chimpanzees seem to be better than at least non-Indigenous humans at paying attention to the behavior of others, and seem to be more sensitive to these others' actions. The article even quotes a behavioral economist from Caltech as acknowledging, "It seems like they're keeping better track of their opponents' previous choices. You can see, compared to the human subjects, they're just more responsive. They're keeping better 'minds' on what their opponents are doing."

Please note his dismissive use of scare quotes around the word "minds." In the "minds" of human supremacists, only humans have minds; the best anyone else can hope for is "minds."

Doesn't the insistence on our separation from all others ever get tiring? Doesn't it all start to seem a little desperate?

The next implication I might draw could have to do with one of the definitions routinely used to declare humans über-intelligent, which is that intelligence is the ability to recognize patterns. But, uh, the chimpanzees are better than we are at recognizing patterns in the play of their opponents, which means, uh, well, maybe we're not number one. Damn it all.

But of course, neither of those are the implications the scientists and journalists draw from all this. The Caltech behavioral economist quoted above concluded, "One theory is that the humans are overthinking it, and the chimps have a simpler model."

Extraordinary. He just turned the fact that chimpanzees outperformed humans in this game into evidence that humans are more complex thinkers. Or maybe it's not so extraordinary. Isn't it what we would expect from narcissists?

This behavioral economist won a MacArthur "Genius" Grant in 2013. So I guess this means that if I want to show myself a more complex thinker than this "genius," I need to play games with him and make sure to lose. Following his "logic," me losing would be evidence I overthought, while him winning would be evidence that his "mind" uses a "simpler model."

Despite, or perhaps because of, the self-serving stupidity of the genius's comment, the editor of the newspaper that printed the article used that comment as a pull-quote.

Of course the editor did.

The conclusions of other human supremacists are equally ridiculous. As the article states, "Researchers believe the different outcomes could be the byproduct of a cognitive trade-off in the course of evolution. Humans left the trees and developed language, semantic thought and cooperation, while our distant cousins kept right on doing what made them so successful in the first place: competing, deceiving and manipulating."[62]

Yes, that's right. They just turned evidence for an increased sensitivity and responsiveness toward their playmates—the chimps were, after all playing a game, which is not quite the same as, say, stealing someone's land and extirpating them, which *someone* we could name has done once or twice or a million times—into evidence that chimpanzees are deceitful and manipulative.

Please note also some of the other propaganda in that paragraph. First, it's irrelevant that humans "left the trees"; how does leaving trees for grasslands imply the development of "language, semantic thought and cooperation"? The phrase "left the trees" pretty clearly is used here as shorthand to signify humans separating themselves—psychologically and spiritually, since of course it's not possible physically—from Nature. Second, nonhumans have highly developed "language, semantic thought and cooperation," which means, much as we humble narcissists like to think we invented everything, that humans didn't "develop" them. Third, it is this culture that is refusing to cooperate with the rest of the world, but is instead projecting its own competitive mindset onto reality (*Selfish Gene*, anyone?). Salmon, forests, and rivers seem to cooperate just fine. The paragraph is really just a recapitulation of the Great Chain of Being, nothing more than the tired re-assertion that "At some point in the past, humans crossed some otherwise impassible chasm that now separates Humans from Nature, stopped being another animal that is red in tooth and claw, stopped being matter, and became mind, became elevated, filled with abstract thoughts (never mind that the chimpanzees were playing this game on a computer, and you can't get much more abstract than that) and became (cue the swell of violins to drown out the screams of this culture's human and nonhuman victims) cooperative."

The original paragraph would be far more accurate if it read, "After some humans metaphorically 'left the trees' by defining themselves as separate from and superior to Nature—and to maintain this self-definition they must put themselves in perpetual *opposition* to Nature—they developed patriarchy, wars of extermination, and ecocide; and they traded cooperation for competition, domination, and manipulation; while these humans' 'distant cousins' have been thrown off their lands, ripped from their families and friends, and subjected to stupid lab tests."

It doesn't really matter whether any of the human supremacist assertions make sense, so long as they serve our sense of superiority. Chimpanzees are better than are humans at these games, which then somehow means humans are superior, smarter, and more cooperative. And besides, chimps are deceitful and manipulative. They must be; it couldn't actually be that they are better than we are at something. The big cheaters. So there.

※ ※ ※

Let's be very clear on what just happened. The human captors devised a game, and when captive chimpanzees beat humans at this human-devised game, the human captors accused the prisoners of being deceitful and manipulative.

And the captors are also claiming that they themselves excel at cooperation. As they hold these others captive.

What a fucking surprise.

❄ ❄ ❄

Doesn't that remind you of when you were in elementary school, and in every grade there inevitably seemed to be this one spoiled kid who invented games with rules that made it so he was always supposed to win, and then whenever someone else would start to win he'd change the rules, then change them again, and when he lost anyway, he'd whine that the other kid must have cheated?

When a child does this, it's unpleasant and pathetic, but sometimes at least understandable in a more-to-be-pitied-than-censured sort of way. When an adult does it, it is very much, as a dear psychologist friend of mine is fond of saying, "diagnostic of something very wrong with the person's emotional and mental health."

❄ ❄ ❄

Not only must human supremacists make the world jump through hoops on command, they must make their own perception of the world jump through hoops on command. This is one manifestation of cutting the vocal cords of the planet. The planet, from this perspective, doesn't really exist. The only reality they can accept is the one they create. Which does not correspond to reality at all.

This is why even when the chimps win, they lose. This is how "knowledge" or "exploration" or "research" works in this culture of human supremacism. Primary purposes of "knowledge" or "exploration" or "research" (as well as, of course, philosophy, religion, ideology, law, and so on) in a supremacist society include increasing the supremacists' beliefs in their own superiority; and even more so, increasing their control over all those they perceive as inferior; which also, not coincidentally, once again increases their beliefs in their own superiority.

Thereby staving off, if only for a little while, the nagging fear that they may not be separate and superior after all.

❊ ❊ ❊

We've all heard of the Milgram experiment, where participants were led to believe they were taking on the role of "teacher" in a study on the relationship between pain and learning. An authority figure told the teacher to administer electric shocks to a "learner" when the learner gave incorrect answers to questions. Unbeknownst to the teacher, the learner—who was in another room and could be heard, but not seen—was in on the experiment, and there were no electric shocks. But as the strength of the "shocks" would increase with each wrong answer, the learner would moan and scream as if in pain, and cry out about his heart condition. The authority figure would push the teacher to deliver ever stronger shocks to the learner. Toward the end the learner might begin banging on the wall, and then go ominously silent.

Most people believed that nearly everyone would stand up to the authority figure and not harm another human being. But most people were wrong: in reality, more than 60 percent of the subjects obeyed the authority figure and tortured the helpless victim to the very end.

Now here's my point: when researchers set up an experiment where a rat received food by pressing a lever, and then added the twist that pressing the lever shocked a rat in a nearby cage, the rat refused to press the lever. Different researchers replicated this experiment with rhesus monkeys, who also refused to torture their fellows. One monkey refused to eat for twelve days, literally starving himself instead of causing another pain.

And who are the cooperative ones?

❊ ❊ ❊

Scientists conducted an experiment in which they starved one capuchin monkey while those in cages nearby were fed (we can certainly ask what sort of sadist would conceptualize such an experiment, but we already know the answer). To their surprise, they found the starved monkey didn't lose any weight. They could only conclude that the other monkeys were surreptitiously sharing their food.

And who are the cooperative ones?

❀ ❀ ❀

Whalers have long known that if they kill or wound one sperm whale, other whales will come to try to help their comrade. The whalers then kill the rest of the pod.

Who are the cooperative ones?

❀ ❀ ❀

Hunters knew that if they killed or wounded one Carolina parakeet, the parakeet's friends would hover around to protect the wounded one. The hunters then killed the rest of the parakeets. In fact they drove them extinct.

Who are the cooperative ones?

❀ ❀ ❀

The other thing that breaks my heart about the plant research is that, unsurprisingly, most of it is done not to help plants, but explicitly to support agriculture or industry: to take without giving back. For example, in the article cited earlier in this chapter about Mother Trees, the author of the article writes, "The concept of symbiotic plant communication has far-reaching implications in both the forestry and agricultural industries. This revelation may change the way we approach harvesting forests, by leaving the older trees intact to foster regrowth. In agriculture, undisturbed mycorrhiza systems enhance plants' ability to resist pathogens, as well as absorb water and nutrients from the soil, bringing into question common practices that disturb these underground networks, such as plowing."

Given the destructiveness of this culture, and given our complete unwillingness to address the depth or insatiability of this destructiveness, I guess we should be glad the author at least acknowledges you can't take *every last tree* from a forest, and that the plow—the invention upon which agriculture (and indeed, the entire culture) is based—might not be particularly benign.

Another article on plant communication segues from asking whether plants communicate or give soliloquies directly to, "The possibility that plants routinely share information isn't just intriguing botany; it could be exploited to improve crop resistance to pests."[63] It's always about exploitation, isn't it? In this case, it's about exploiting this newfound human understanding of plant language specifically so members of the dominant culture can more efficiently exploit the plants they call crops. This is standard behavior by those who conquer: learn enough of the local languages to facilitate enslavement and exploitation of those whose land they've occupied.

Or there's this: "Research on plant communication may someday benefit farmers and their crops. Plant-distress chemicals could be used to prime plant defenses, reducing the need [sic] for pesticides. Jack Schultz, a chemical ecologist at the University of Missouri, who did some of the pioneering work on plant signaling in the early nineteen-eighties, is helping to develop a mechanical 'nose' that, attached to a tractor and driven through a field, could help farmers identify plants under insect attack,[64] allowing them to spray pesticides only when and where they are needed [sic]."[65] Always more efficient ways to exploit, not to know or relate or help any others on their own terms.

Even someone who loves plants as much as does Stefano Mancuso, someone who has devoted his life to understanding and helping them, is not immune to the highly contagious mental illness that is human supremacism. Michael Pollan writes, and I quote this at length (broken up by my responses) because it is, to me at least, so completely heartbreaking and horrifying: "If we could begin to understand plants on their own terms, he [Mancuso] said, 'it would be like being in contact with an alien culture. But we could have all the advantages of that contact without any of the problems—because it doesn't want to destroy us!'"

Well, I think that at this point if we were to begin to understand plants on their own terms, the first thing they would tell us is to stop enslaving everyone, and to stop murdering the planet. If plants send chemical messages letting their neighbors know their leaves are being eaten by caterpillars, what chemical messages might they send when entire forests, including Mother Trees, are clearcut, when marshes are drained and paved, and when grasslands are plowed under and planted to corn?

I'm sure from the perspective of plants, *we* are the "aliens" who want to destroy them. If we understood plants from their perspective, we would know this.

Within the context of this exploitative and destructive culture, what Mancuso's comment really means, especially in practice, is that we would gain tremendous advantages from this "understanding" not because the plants "don't want to destroy us," but far more accurately, because they don't fight back as we exploit and exterminate them; they continue, to use Dawkins's term, to be "Suckers." From this perspective, the hope is that our understanding of plants will allow us to become more effective Cheats, to more effectively and with ever greater impunity steal from them.

Finally, if space aliens *did* conquer the earth, we all know what their relationship to human languages would be. At first they would deny that humans have language, and then when they finally did allow that possibility, they would learn our languages specifically so they could use that knowledge to facilitate our further enslavement and exploitation. Sound familiar?

Pollan continues, "How do plants do all the amazing things they do without brains? Without locomotion? By focusing on the otherness of plants rather than on their likeness, Mancuso suggested, we stand to learn valuable things and develop important new technologies. This was to be the theme of his presentation to the conference, the following morning, on what he called 'bioinspiration.' How might the example of plant intelligence help us design better computers, or robots, or networks?"

Really? That's why you want to learn more about plants? So you can help humans to more effectively dominate the natural world?

Imagine for a moment that we're living in the alien contact scenario Mancuso mentioned above. In this scenario the aliens really do want to enslave, exploit, and destroy us. In fact, the drive to enslave, exploit, and destroy us is so strong among these aliens that even the most gentle and kind of them—even those who genuinely love us (insofar as these bizarre aliens are capable of what we humans would recognize as love)—attempt to learn about our physiology, our languages, our relationships, not so they can help us resist the alien exploitation, nor so we can better be left alone, nor even out of simple curiosity, but instead so they can learn how our skin and bones and bodies and minds and organs are made up, so they can design ever more efficient alien computers, robots, and networks, so they can make ever better ways to exploit us, to enslave us, to make us jump through hoops on command, so these aliens can attempt to come closer to taking complete control of our lives and turning every bit of the planet to alien use, and in the process, destroying life on the planet.

And they will call that getting to know our perspective.

Pollan continues, "Mancuso was about to begin a collaboration with a prominent computer scientist to design a plant-based computer, modeled on the distributed computing performed by thousands of roots processing a vast number of environmental variables."

These aliens will dissect our brains, trying to figure out how we think, so they can design human-based computers (after all, some humans claim human brains are the most complex phenomenon in the universe) that will facilitate alien commerce and industry, and ultimately alien control of the planet.

"His collaborator, Andrew Adamatzky, the director of the International Center of Unconventional Computing, at the University of the West of England, has worked extensively with slime molds, harnessing their maze-navigating and computational abilities. (Adamatzky's slime molds, which are a kind of amoeba,[66] grow in the direction of multiple food sources simultaneously, usually oat flakes, in the process computing and remembering the shortest distance between any two of them; he has used these organisms to model transportation networks.) In an e-mail, Adamatzky said that, as a substrate for biological computing, plants offered both advantages and disadvantages over slime molds. 'Plants are more robust,' he wrote, and 'can keep their shape for a very long time,' although they are slower-growing and lack the flexibility of slime molds. But because plants are already 'analog electrical computers,' trafficking in electrical inputs and outputs, he is hopeful that he and Mancuso will be able to harness them for computational tasks."

One of the alien scientists commented to an alien reporter that humans have many advantages over slime molds, in terms of turning these humans into what he called "living computers." He said, "Humans are more robust, and can keep their shape for a very long time, although they are slower-growing and lack the flexibility of slime molds." When asked whether turning these humans into living computers might cause them distress, the alien laughed and said, "Of course not. No alien brain, no pain." The alien also mentioned that in order to maintain quiet workplaces, the aliens normally cut the vocal cords of the human-computers. "Not that these vocalizations are indicative of any sort of primitive distress signal," he said. "It just makes for a safer and more stable work environment."

Pollan continues, "Mancuso was also working with Barbara Mazzolai, a biologist-turned-engineer at the Italian Institute of Technology, in Genoa, to design what he called a 'plantoid': a robot designed on plant principles. 'If you look at the history of robots, they are always based on animals—they are humanoids or insectoids. If you want something swimming, you look at a fish. But what about imitating plants instead? What would that allow you to do? Explore the soil!' With a grant from the European Union's Future and Emerging Technologies program, their team is developing a 'robotic root' that, using plastics that can elongate and then harden, will be able to slowly penetrate the soil, sense conditions, and alter its trajectory accordingly. 'If you want to explore other planets, the best thing is to send plantoids.'"[67]

The aliens continued, "If you want to explore other planets, the best thing is to send humanoids, which are robots made like humans." When asked whether these humanoids would have intact vocal cords, the alien again laughed, and said, "I can't see any reason why not. In space, no one can hear you scream."

* * *

Since I love so much of Stefano Mancuso's work, my fervent hope is that secretly he's as horrified as I am at some of the abuses of plants toward which some of his research aims, but he's lending his talents and his name toward these destructive ends because he knows that's the only way he can get grants, out of which he'll be able to also pay for the projects he really wants to fund, you know, the ones that help plants.

Basically, I guess I'd rather it be the case that he's making heartbreaking compromises (recognizing that plants pay the real cost of these compromises) so he can also use his voice as a powerful advocate for plants, instead of it being that he's consciously or unconsciously attempting to divert the increasing and inevitable acknowledgment of plant intelligence back into the culturally acceptable realm of human supremacism.

I realize that in both cases the same research gets done (some of it liberating, some of it furthering enslavement), but the former case seems disheartening to me, and points toward the struggles and compromises many make within a capitalist system, performing destructive activities to raise money in order to do other activities that help those they're harming when they work for pay.

And the latter case seems disheartening to me in a different, yet also altogether all-too-familiar way, as one of this culture's ways of dealing with liberating ideas that are gaining enough recognition that they can no longer be ignored or ridiculed or crushed out of awareness, is to vigorously co-opt the ideas back into the service of existing bigotries and hierarchies. This happens all the time, as, for example, the movements for breaking the hegemony of Eurocentric and masculine stories through multiculturism got diffused and dead-ended into postmodern relativism, and the movements for women's liberation and the rights of gays and lesbians got co-opted into servicing patriarchy through the misogynist mess that is queer theory, and environmentalism was transformed along the way from an attempt to save the real, physical world from biocidal industrial civilization toward attempts to "sustain" precisely the civilization that is destroying the world that the new "environmentalists" pretend they want to save (i.e., environmentalism has gone from being about saving wild places and beings toward promoting, for example, wind energy; just tonight I saw a dreadful interview with a so-called environmental publicist who was saying that environmentalists need to never use the words "Earth" or "planet," and instead only talk about what they will do to "improve" human lives; or there's this gem from The Nature [sic] Conservancy's chief scientist, Peter Kareiva: "Instead of pursuing the protection of biodiversity for biodiversity's sake, a new conservation should seek to enhance those natural systems that benefit the widest number of people. . . . Conservation will measure its achievement in large part by its relevance to people.").[68] This culture excels at that co-optation, and I hope that the burgeoning understanding of plant intelligence isn't being sucked into the service of atrocious ends, as this culture does to so many other movements, ideas, and ideals.

❀ ❀ ❀

There's a third possibility, and this is the one I find most likely by far, which is that Mancuso, like pretty much all of us in this culture, has so fully internalized this culture's unquestioned human supremacism that the dissonance between his love of plants and the exploitation of them inherent in some of this research never rises fully into consciousness. Or at least not enough to impinge on the research. This is pretty much the definition of an unquestioned belief.

❀ ❀ ❀

He's got his own human supremacist blind spots. I've got mine (although by definition, I can't see what they are). And you've got yours.

❀ ❀ ❀

The point here isn't Stefano Mancuso. The problems are human supremacism and a system that socially rewards practices that harm nonhuman communities. This can lead people to take morally contradictory stances. For example, earlier I cited the scientist Anthony Trewavas, who is a fierce and unstinting advocate for a recognition of plant intelligence. Yet at the same time he is perhaps even fiercer and more unstinting in his advocacy for genetic modification and biotechnology, and in his criticism of organic farming (for standard industrial and pro-corporate reasons).

❀ ❀ ❀

One of the myths of modern culture is that science is value free. That's nonsense, of course. Not only because reality is necessarily more complex than *any* analysis or interpretation of reality, which means that by definition, values must be imposed through what is and is not included in the analysis or interpretation; and not only because, protestations of some humans aside, the universe is far more complex than a human brain (and of course far more complex than a computer), and is far more complex than we are capable of thinking (and of course far more complex than machines are capable of computing). This myth of value-free science is only tenable if you've forgotten that unquestioned beliefs are the real authorities of any culture, and then if you presume that anything that questions those assumptions is "speculation" or "philosophizing" (as opposed to those more legitimate "analyses" that fail to question the assumptions).

It's pretty funny, really, or would be if it weren't killing the planet. At least some of us some of the time understand that science performed by those who work, for example, for Monsanto, may very well be tainted by self-interest (or in this case the interest of the corporation, which, because, as Upton Sinclair said, "It's hard to make a man understand something when his job depends on him not understanding it," takes us right back to a skewed sense of self-interest). At least some of us at least some of the time might laugh at the science performed by those who worked for tobacco companies in the middle of the twentieth century that purported to show that tobacco wasn't harmful. Science is supposed to be "disinterested," we always hear. And we always hear that if it's not "disinterested," then it's not science. Never mind that, "'For better or worse,' said Steven A. Edwards, a policy analyst at the American Association for the Advancement of Science, 'the practice of science in the 21st century is becoming shaped less by national priorities or by peer-review groups and more by the particular preferences of individuals with huge amounts of money.'"[69] I'm not sure how new this is; industry and industrialists (and the military-industrial complex) have driven science from the beginning. It is only because we want to forget all this that we pretend science is disinterested in the first place.

But even without this obvious conflict of interest, this is all just the same old shit we've been seeing all along: if you think plants communicate, you're speculating, and if you think plants learn, you can't get your work published; but if you want to torture plants into living computers, then you get a fucking grant, and if you figure out how to violate their very genes, you'll win a fucking prize.

Oh yeah, I guess I forgot: science is value free.

※ ※ ※

Here's the point: research that in some way or another attempts to extend human control over the universe is considered value free. Hell, attempts by humans to control the universe—to make matter and energy jump through hoops on command, and to predict what will happen and when—are in our very definition of how we know something is true.

But extending human control over the universe is a value. And it is a value that materially benefits (in the short term, so long as you don't mind a murdered planet) the humans doing the research, and those funding the research, and those publishing the research, and those using the technologies that emerge from the research. They are not and can never be "disinterested."

If a white person does research that facilitates the enslavement of members of other races, at least some of us would recognize the very real possibility that the white person's "research" does not in fact represent reality, but instead has been skewed to rationalize the enslavement. How is it that when it comes to the enslavement of nature—which includes as much of the universe as we can manipulate—we suddenly get really stupid?

Our stupidity has the same source as would the racist researcher's in the previous paragraph. This time we'll misquote Upton Sinclair: "It's hard to make a man understand something when his entitlement depends on him not understanding it." Our stupidity in this case is an inevitable consequence of, and inevitable defense of, our human supremacism. It's an inevitable consequence of a naturalistic [sic] philosophy that holds only human functionality to be true functionality, and only human (and in fact scientific) intelligence to be true intelligence. Of *course* attempting to extend human domination over as much of the universe as possible is seen as either value neutral or positive, no matter how much this attempted domination harms the real, physical world. Humans (and in fact industrial humans) are the only ones who matter. Industrial humans are the only ones who exist. Human functionality is real. Functionality in the real, physical world is not real functionality. A river serves no purpose till it is harnessed for electricity, transportation, and irrigation. A forest serves no purpose till it is converted into 2x4s.

You'd think that when unquestioned assumptions are the real authorities of a culture, and when this culture is killing the planet, that it might be long past time we questioned some assumptions, and long past time we questioned some values, and long past time we questioned what we perceive as true.

CHAPTER SIX

WONDER

I would feel more optimistic about a bright future for man if he spent less time proving that he can outwit Nature and more time tasting her sweetness and respecting her seniority.

E.B. WHITE

Slime molds are pretty cool, for a number of reasons.

First, they used to be classified as fungi, but recently were reclassified as amoebas. This gives me hope that someday some sapient classifiers may reconsider the whole *Homo sapiens sapiens* thing.

We've already talked about the next cool thing: slime mold's ability to learn and remember. But there's more coolness ahead.

Before we get to that, though, we should probably mention who they are. They're tiny beings who feed on microorganisms like bacteria, yeasts, and fungi who live in dead plant material. One of their gifts to the larger community is that they can contribute to the decomposition of dead vegetation. When this food is abundant, they live independently as single-celled organisms. But, and here's where it gets even cooler, when food is less common, these single-celled beings can join together and begin to move as one, often following scents toward new food sources. The individuals can change their shape and become different functional parts of this collective; for example, they can become a stalk that produces fruiting bodies that release spores.

Yes, that says what you think it does. They can transition from single- to multi-celled creatures. And then move as one. And they can morph!

This is precisely the opposite behavior of that predicted in many models proposed by mechanistic scientists. In the "run on the bank" model, and in the similar "grocery store running out of food" model, so long as there is plenty of money in the bank (or food in the grocery store), people are polite. They will wait in line. They will observe social niceties. But when resources become scarce, people push and shove their way to the front of the line. They lie and cheat and steal. They do not act communally. They in fact act anti-communally. But slime molds act precisely the opposite of what these models predict: when the going gets tough, slime molds recognize the importance of community.

❊ ❊ ❊

I want to mention the single stupidest argument I've seen against plant intelligence. It's from an essay on the website of a scientific philosopher. Why doesn't that surprise me? The essay first mentions that some scientists understood the existence of "plant signaling" (i.e., plant communication) as long ago as 1935, and then goes on to say, "If chemical signaling in plants warrants re-evaluation of our moral attitudes towards plants, then such a re-evaluation would have been appropriate in 1935. But it wasn't appropriate in 1935, so chemical signaling shouldn't warrant any change in ethical attitudes now."[70]

Gosh, I can't think of any moral attitudes from 1935 that warrant reevaluation. Well, except for maybe that thing about the moral attitudes of Nazis against Jews. Or the moral attitudes of whites against members of other races. Or the moral attitudes of men against women. Straights against gays and lesbians. The civilized against Indigenous peoples. This culture's contempt for the natural world, indeed, its hatred. And so on.

And just so we're clear, this was not on the website of some undergraduate philosophy student who is filled with that unbeatable combination of ignorance and certainty that for the most part only seems possible for those between the ages of seventeen and twenty-five. The website is run by the Chair of the Department of Philosophy at a college of the City University of New York and the editor of a journal of scientific philosophy.

❊ ❊ ❊

Tonight I watched an old episode of *QI*, a British quiz show where the host asks strange questions of a panel of comedians, rewarding interesting and correct answers and taking away points (and making a general hullabaloo) when contestants give answers that are both boring and incorrect.[71]

One of the strange questions tonight: "If aliens arrived on earth to abduct our most successful inhabitant, where would they look?"

After some jokes came the host's response: "By any criteria by which you judge the success of life, bacteria win hands down in terms of profusion, diversity, ability to live under extraordinary conditions. . . . We wouldn't be alive without them. We entirely depend upon them. . . . If you were to take a gram of soil, there are 40,000 species in that one gram. And each species is as different from each other as a rhinoceros is from a primrose. I want you to fall in love with the bacteria. They are the most marvelous things conceivable. They live in boiling acid, they live in ice, they live in nuclear cooling water. They can live absolutely anywhere, for example under six thousand atmospheres of pressure. They love the human tummy. We reckon that 75 percent of bacteria in the human tummy have not yet been identified by species. They're fantastic."[72]

✳ ✳ ✳

Humans are superior and special because we're so adaptable? Humans ain't got nothin' on bacteria.

✳ ✳ ✳

I'd also add that bacteria essentially made life on this planet. Without bacteria there would be no life here. Without humans, life would go on very well, thank you very much. In fact, given how *Homo sapiens sapiens* are acting, there is some doubt as to whether life on this planet will continue.

Now who is superior?

There's that question that always resides at the core of this culture's violation imperative and ramifies into its outer reaches: if bacteria can create life on this planet, and humans are doing their damnedest to destroy it, who, then, is stronger, the creator or destroyer?

This question, as absurd as it seems on the surface, is in many ways key to understanding this culture's self-described superiority, its destructiveness, and its perception of its own destructiveness as a sign of its superiority.

※ ※ ※

Bacteria communicate. If one group of bacteria gains resistance to antibiotics, they can help others to gain resistance, too. An article in *Science Daily* entitled "Bacteria Communicate to Help Each Other Resist Antibiotics," states, "The more-antibiotic-resistant cells within a bacterial population produce and share small molecules with less-resistant cells, making them more resistant to antibiotic killing." The first author of the article, Omar El-Halfawy, notes, "These small molecules can be utilized and produced by almost all bacteria with limited exceptions, so we can regard these small molecules as a universal language that can be understood by most bacteria."[73]

Antibiotics, however, are by no means the only things bacteria talk about. As a writer for *Scientific American* put it, "Forty years ago scientists discovered that some bacteria send and receive messages—in the form of small molecules—to and from surrounding cells. This kind of communication, called quorum sensing, enables bacteria to monitor their population density and to modulate their behavior accordingly. When there are enough cells around to create a 'quorum,' bacteria begin producing proteins known as virulence factors that sicken their hosts. They can also grow into aggregates called biofilms that render them up to 1,000 times more resistant to antibiotics."

I understood the part about biofilms, but not the part before that. How is it in the best interest of these bacteria to sicken the hosts when they gain enough cells to create a "quorum"? Why would they do that?

I asked a friend who got his PhD working with biofilms.

He responded, "The article from *SciAm* is actually an oversimplification (as is almost all science writing for public consumption). The truth is that quorum sensing will do different things depending on the bacteria in question and its lifestyle. Most bacteria, which are not pathogenic, would use QS to regulate some aspect of their lifestyle. Some organisms will use it as a way to turn on virulence factors, while others use it to turn them off.

For example, a bacteria that infects via the fecal-oral route (we all eat small amounts of feces all the time, sometimes it carries disease) and causes an acute gastroenteritis might want to turn on genes that code for a toxin to stimulate diarrhea to boost its transmission. *Staphylococcus aureus*, however, will often switch to sessile biofilm mode of growth, repress virulence factors, and set up shop somewhere in your body like a heart valve or the bone (causing endocarditis or osteomyelitis respectively).

"The takeaway message is actually much broader—bacteria dynamically react to stimuli to change a wide range of host behaviors from virulence factor production, changing movement speed, altering metabolism, etc. Bacteria can sense a wide range of things, including nutrients, ions, and temperature; and can change to adapt to their environment and (if they are pathogens) ensure their transmission."

Well, that makes a lot more sense, and frankly makes bacteria all the more impressive to me.

Let's return to the *Scientific American* article, and its next sentence: "Quorum sensing is now known to be widespread in the bacterial world, and many researchers hope to develop ways to disrupt it."[74]

Okay, so you know that bacteria are foundational to life on earth; and you know that in fact in your own body, cells of bacteria outnumber human cells by about ten to one;[75] and you know that bacteria are everywhere on the planet, which means they're effectively impossible to quarantine on a large scale; and you know that bacteria communicate; and yet you "hope to develop ways to disrupt" one of their "widespread" functions? Gosh, what could possibly go wrong?

❊ ❊ ❊

One more thing: if aliens *were* to invade the earth, another reason they'd try to learn our languages would be to try "to develop ways to disrupt" our resistance toward them.

❊ ❊ ❊

Slime molds aren't unique in their ability to transform from single-celled to multi-celled creatures. An article entitled "Future research trends in the major chemical language of bacteria" states, "The discovery of chemical communication among bacteria revolutionized the thinking that bacteria exist in isolation as single-celled organisms. It has become evident in the last fifteen years that bacteria have the potential to establish highly complex and often multispecies communities." These bacteria can then participate in "coordinated and synchronized community behavior."[76]

* * *

I keep thinking about bacteria communicating antibiotic resistance to each other, including those quite unlike them. And I keep thinking about that question asked by the scientists about why plants would help "competitors" to resist some danger. And I keep thinking about the narcissism of this culture, and how both narcissism and supremacism keep making us ask the wrong questions. I don't think the question is, why would bacteria "waste" energy "clueing" in "competitors" about some danger. I think these are better questions: Why do we keep trying to hold ourselves separate from everyone else? Why do we keep trying to believe we are superior to them?

* * *

The neuroscientist John Allman has said, of bacteria, "Some of the most fundamental features of brains, such as sensory integration, memory, decision-making, and the control of behaviour, can all be found in these simple organisms."[77]

* * *

Bacteria are also faster than we are, relative to their size. Most bacteria can move ten times their body length per second. Some can move 100 times their body length. The fastest humans can run about five times their body length per second.

Damn it, we aren't even superior in that way.

❊ ❊ ❊

I know what the human supremacists are going to say. Dude, anytime you drive down the interstate, you're probably going ninety or 100 feet per second. That's sixteen of your lengths per second. Oops. Still not as fast as some bacteria. But when you get on a jet you're going about 900 feet per second, which is 150 times your body length. And the fastest jet went more than 4,500 miles per hour, which is more than 6,600 feet per second, which is 1,100 times your body length per second. Score one for the humans. You can't talk about humans—and human superiority—without talking about technology.

Never mind that bacteria are on the plane, too, which means they're going even faster relative to their body size than humans.

Oh, well.

I'm not denying that humans can go fast. But don't you think a relevant question might include, at what cost?

What infrastructures are required for humans to go even sixteen body lengths per second, much less 160 or 1,100? What are the ecological costs (ignoring, for now, the social and psychological costs) of these infrastructures? Who pays these costs?

I don't think at this point I have to detail the costs of the oil economy (or, for that matter, the industrial economy, or civilization itself) on individual nonhumans, on nonhuman communities, or on the planet (leaving aside for now its costs on humans in the colonies). It should be clear to anyone paying any attention whatsoever that the oil economy, the industrial economy, and civilization have all been complete disasters for the natural world.

The usual next argument by human supremacists is that these costs are worth it. But the problem is that those foisting costs onto others don't get to decide if it's worth it. Of course the ones who are privatizing profits and externalizing costs are going to say that the profits more than make up for the costs.

The thing I don't understand is how people who make this argument somehow also try to claim superior intelligence. How's this: why don't you use your money to buy a car that costs $25,000, and then I'll take it and sell it for $10,000 and keep the profits? Then we can do the same thing tomorrow, and the next day. What a deal! Pretty soon I'll be rich, and then we'll know for sure that I'm superior!

Or hell, let's leave the car out of it. You (the world) deposit wealth (trees, fish, minerals, soil, and so on) in a bank (the world) and everyone in the world lives off the interest. In fact each year there is more wealth than the year before. This is because every tree, every fish, every living being (which means every being) deposits more wealth than it takes. How else do you think the world became so wealthy? By everyone making the world rich by living and dying. Everybody wins! Except I don't like the arrangement. I want it all. So you deposit wealth and I find ever-more-sophisticated ways to steal from you and from everyone else. This means I'm a genius! I'm superior. It is settled. It's time for me to break into my favorite song: "No time for losers, 'cause I am the champion of the world!"

❀ ❀ ❀

Another question we need to ask about humans traveling so fast is, for how long?

Even before the invention of automobiles or airplanes, humans were still capable of traveling at up to 200 feet per second, or thirty to forty times our length per second.

Just not for very long. Two hundred feet per second is the terminal velocity reached when a human jumps off a cliff.

Of course going this fast is shortly followed by terminal catastrophe.

I'm sure you can see the metaphor, right? For how many species has the infrastructure necessary for humans to travel at current speeds already been a terminal catastrophe, and for how many decades total will humans have been able to travel at these speeds before terminal global catastrophe? What are the causal relationships between humans going this speed and global catastrophe?

❀ ❀ ❀

I've got more bad news for the human supremacist crowd, which is that there's a sense also in which bacteria can be considered to be immortal.

Yes, immortal, that wet dream of monotheists, technotopians, and other human supremacists, who all seem to take umbrage at the fact that we, like everybody else, must die.

Or I guess it'd be more accurate to say, like *almost* everybody else.

Here's how it works. Bacteria reproduce by dividing. Each daughter is the same as her mother—in fact, is *part* of her mother—except that whatever molecules inside her that had become damaged are now diluted in her offspring, and then diluted again in theirs, with this dilution acting as a form of self-rejuvenation. Of course individual bacterium die all the time, just like almost anybody else, but bacteria could still easily claim immortality. Here's why. Because they reproduce by division, as opposed to sex, where a new being is created by the combination of cells from multiple parents, with bacteria there's an unbroken line of self-ness in each one going as far back as, well, the beginning. Moreover, you could argue that the death of one bacterium is the death of only one small part of that bacterial self.

❊ ❊ ❊

Bacteria aren't the only beings who are in some sense immortal. There are water bears, hydras, a certain sort of jellyfish, and lobsters. Yeah, I didn't believe it either. But it's true. And we may as well add bdelloids, planarians, and turtles to the list. And although glass sponges aren't exactly immortal, they can live for more than ten thousand years.

Water bears have a somewhat different form of immortality than bacteria. Water bears, also called tardigrades, also known as moss piglets (and you thought their name couldn't get any cuter than *water bears*, didn't you?), are tiny (half-millimeter) creatures classified sort of half-way between arthropods and nematodes. As you can tell from their names, they like to live in water or in moist places like moss, and they look like little eight-legged bears or piglets. They're as cute as their name. And they're *everywhere*. They've been found on the tops of the highest mountains, and in hot springs, and in ocean sediments, and under layers of solid ice. They are also found in meadows and lakes, and in stone walls and roofs.

Like I said, water bears aren't "immortal" in the same sense as are bacteria. In fact their normal life span might be about a year. Their "immortality" in this case is more like "extremely tough." We've all heard the story of how hard it was to kill Grigori Rasputin, right? According to the story, he was poisoned with enough cyanide to kill a regiment, stabbed, shot, beaten, left for dead, and when the killers came back he lunged at them, so they wrapped him in chains and threw him into a river. When his body was recovered he had water in his lungs, which means the cause of death was said to be drowning. It ends up, as so often happens, that the reality was far more prosaic: he wasn't poisoned, and after being beaten, he was killed instantly with a gunshot to the middle of his forehead. Be that as it may, water bears make even the mythical Rasputin seem as fragile as a Facebook friendship.

Water bears are normally fairly soft and squishy, kind of like itsy-bitsy caterpillars. But if we can say that when the going gets tough slime mold get communal, when the going gets tough water bears desiccate and hibernate. They can reduce the water content of their bodies to between one and three percent. In this state they're virtually indestructible. They can survive a temperature range of 800 degrees, from 350 degrees above zero to more than four hundred and fifty below: they can survive a few minutes at only two degrees above absolute zero. They can survive pressures from the vacuum of outer space to six times the pressure at the bottom of the Marianas Trench (6000 atmospheres). They can survive radiation levels 1000 times higher than those that would kill any other animal. They can survive many toxic chemicals. They can live without food or water for ten years, only to rehydrate, forage, and reproduce when times get better.[78]

But I thought they only lived about a year. That's the thing: including hibernation, they can easily live sixty years. Some estimates say they can live up to two hundred. I know, a lot of that might be spent hibernating, but we still say humans live seventy years, and we spend at least twenty of that sleeping, so I don't think we have much room to complain at saying they can live for decades, if not centuries.

I want to mention one more thing about water bears before we move on, which is that while humans know that these creatures survive some of these extreme conditions because water bears are found living in them, we know they survive others because humans have intentionally put them, for example, into outer space, or into extreme temperatures, or under intense radiation, simply to see if these others can survive. Research like this always leads me to ask: what sort of fiend would do this? What sort of monster would take a creature from its home and send it into outer space, or drop it to almost absolute zero, just to see how long it will survive? What would we think of aliens who did this to humans? What do we think of Nazis who did this to humans? Just last night I was reading about the Nazi Sigmund Rascher, infamous for his vivisection of Holocaust victims. He conducted experiments in which he subjected prisoners to extremely rapid de- and re-pressurization, ostensibly to help fighter pilots who might have to bail out at high altitude; experiments in which he exposed prisoners to freezing water and then used different methods to restore their body warmth; and experiments in which he shot or otherwise inflicted horrible injuries on prisoners, then checked how well new drugs slowed the bleeding. So, what would you think of Nazis, or space aliens, or anyone else, who sent your friends or loved ones, or you, into extreme conditions, just to satisfy their curiosity as to how much you can survive?

This is human supremacism. It's also sadism.

Also, the next time any human supremacists comment on how extraordinarily resilient and adaptable humans are, water bears may simply laugh and say, "Space suits? We don't need no stinking space suits."

❧ ❧ ❧

Bdelloids are "immortal" in the same sense as are water bears, in that they are creatures—rotifers—who live in water, and who can voluntarily enter states of long-term hibernation in which they can survive extreme conditions, then voluntarily wake up, perhaps give themselves a good shake like a pup getting up from a nice nap, and continue with their lives. But they also do something else very interesting. They generally reproduce asexually, which of course many other beings do. But one potential problem of asexual reproduction is that when your genetic material gets damaged in the daily rough-and-tumble of living, those flaws can be passed on intact to your children. If you recall, bacteria deal with this difficulty by exponentially diluting the flaws into essential meaninglessness.

Bdelloids use another means: they beg, borrow, or steal DNA from other creatures. I'm going to let Traci Watson of *Science Magazine* describe it: "In Mother Nature's edition of the TV reality show *Survivor*, the bdelloid rotifers would probably be the last animals standing. These tiny aquatic creatures can survive high blasts of radiation and years of desiccation— and they've persisted for tens of millions of years without sex. Now, a study published online today in *PLOS Genetics* hints at how the bdelloids do it. A new genetic analysis shows that roughly 10% of the bdelloids' active genes were pilfered from other species, such as fungi, bacteria, and plants. These foreign genes have endowed bdelloids with talents that no other animal can boast, which could help explain their ability to shrug off extreme conditions of aridity. Ultimately, the bdelloids' appropriation of foreign genes may hold the key to their success despite celibacy, which usually results in a species's extinction."

Watson also says, "For creatures of such superherolike ability, microscopic bdelloids—which [sic] are distantly related to flatworms—are happy in humble surroundings. The roughly 400 species of bdelloids live in fresh and brackish water, including puddles, sewage-treatment tanks, and drops of moisture adhering to soil. They have a handy ability to survive the sudden disappearance of their aquatic homes; the desiccation-survival record is nine years.

"What's even stranger, from an evolutionary biologist's point of view, is the bdelloid's long-term asexuality. For perhaps 80 million years, all bdelloids have been shes, contentedly reproducing without males—and defying biologists' ideas about the centrality of sex. Sexual reproduction, the thinking goes, introduces genetic variation and so allows a species to adapt to a changing environment and to genetic degradation. It's commonly thought that animals that [sic] forgo sex eventually go extinct, but the bdelloid provides a glaring exception to the rule. Legendary biologist John Maynard Smith was so flummoxed by the bdelloids that he called them an 'evolutionary scandal.'

"In 2008, a separate group of researchers found that bdelloids contain some foreign DNA in a small region of their genomes. Tunnacliffe and his colleagues decided to find out the extent of that foreign genetic material. So they turned to the bdelloid *Adineta ricciae*, which was discovered in a small Australian billabong, or lake. When the scientists sequenced the bdelloid DNA that provides the blueprints for active genes, they found that roughly 10% of that DNA had been borrowed from some other creature. All told, the bdelloid had adopted DNA from more than 500 different species.

"By comparing the foreign sequences to genetic databases, the researchers learned that many of the sequences are responsible for directing the production of enzymes found in simple organisms but unknown in complex animals. Two genes, for example, give rise to bacterial enzymes that help break down the toxic chemical benzyl cyanide. Two more genes, these from parasitic protozoa, direct the manufacture of a compound that can ward off cellular damage. Nearly 40% of the animal's enzymatic activity includes a foreign component, Tunnacliffe says."[79]

I get so tired of human supremacists, who perceive us as superior because we can make toasters, protein shakes, and unicycles. Every being has its own gifts.

The next "immortal" creatures are planarians, who are tiny, non-parasitic flatworms. Like the others we're mentioning, they can certainly die. But part of their "immortality" claim comes from their ability to regenerate. If you cut off a flatworm's head, the body will grow a new head and the head will grow a new body. If you cut it down the middle, it will grow into two flatworms. If you cut a flatworm into pieces as small as 1/279th of the original body size, the pieces will regenerate into new flatworms.

Another part of their "immortality" claim comes from being, well, immortal. Basically, part of the reason we age is that as our cells reproduce, the ends of the chromosomes can become, to use a completely non-scientific term, frayed. Or, to be more precise, whenever your chromosomes reproduce, there are reasons that are functional to this reproduction that make it so the chromosome cannot be duplicated to the very end. The (highly intelligent, in my estimation) response by bodies to this problem is to create "telomeres," which some people call "disposable buffers," at the ends of chromosomes. Each time the chromosome is duplicated, you lose a little of the telomeres, which means the telomeres can only protect the chromosomes (and you) for so long. Once the chromosomes themselves start suffering damage, the cells can no longer properly reproduce. In practical terms, what this means for humans is that as we get older, we can no longer run as fast as we used to, we can't heal so quickly, we have less supple skin, we lose interest in sex, perhaps we gain an interest in either Bingo or watching *Judge Judy*, and eventually we suffer organ failure and death (which might, now that I think about it, have been caused by watching *Judge Judy*). Creatures who reproduce sexually deal with this reality by having babies, who have brand spanking new chromosomes, at some point after which the parents—and their aging, damaged chromosomes—die. Bacteria deal with this reality, as we've said, by dilution. Well, when you cut a planarian, the part that was cut off is replaced with cells that are as good as new, in part because they actually *are* new. As long ago as 1814, this caused one scientist to say that flatworms can "almost be called immortal under the edge of the knife."[80]

But flatworms are even more amazing than this. If you teach flatworms to be frightened of bright lights, by flashing lights just before you shock them, and then you cut these flatworms in two, the resulting flatworms are still terrified of bright lights. That of course makes sense, because these are the same flatworms you already traumatized. It also means that these regenerated worms remember their experiences from before, which means, as with bacteria, that there can be an unbroken line of self-ness and memory back to the beginning of that string of regenerated flatworm.

If you grind up these flatworms and feed the mash to more flatworms, the new flatworms learn to associate bright lights with shocks far faster than would be otherwise expected. The new flatworms somehow metabolize or take in the memories of the flatworms they ate.

When I read that, I wondered at the complexity of the real world and also what sort of mindset might lead someone to not only terrorize and torture someone, but then come up with the idea of killing the beings he'd previously terrorized, grinding them up, and feeding them to a new batch of victims.

The next creature I want to talk about are hydras, who are immortal in the old-fashioned sense that they don't age, and they don't die of old age.

Hydras are small—from a few millimeters up to a maximum of about thirty millimeters long—tube-shaped fresh water animals. Picture tiny versions of their close relatives, the sea anemones: a barrel-shaped body with a mouth on top surrounded by stinging tentacles. And in both cases the tentacles are used for the same reasons: sea anemones use them to paralyze and kill fish, then bring the fish to their mouths, and hydras do the same to microscopic invertebrates like daphnia or cyclops.

Hydras can either reproduce asexually, by "budding" baby hydras out of their body walls (in which case each baby is identical to its parent); or sexually, with females growing eggs in their body walls, then hoping that sperm released into the water by males will encounter their eggs and together create little hydra bundles of joy.

Frankly, our means of conception sounds more fun.

But now that I think of it, how do we have the information to disparage the lovemaking of other species? How do we know what ecstasy hydras do or do not feel at the intercourse of body and water and water and body? How do we know that not every being—plant, fungi, animal, rock, river, mountain, virus, bacteria—perceives its own method of reproduction as the best or most pleasurable or fun? How do we know that bacteria do not feel sorry for those who do not divide, and huckleberries do not feel bad for those whose love does not include ecstatic associations with pollinators?

Back to hydras. Hydras do not age. Their stem cells have a capacity for more or less perpetual renewal. Or, to use a term created by Caleb Finch, gerontologist and author of *The Biology of Human Longevity*, hydras have "negligible sensescence." They have no measurable reductions in reproductive capacity with age. They have no measurable functional decline in strength, mobility, and so on, with age. They have no increased death rate with age.[81]

Once again, they can still be eaten or get some disease, or they can be captured by scientists and cut up into small pieces.

Oops. That last one won't kill them. Like planarians, hydras can, when cut up, regenerate. Not only that, if put through a sieve, the cells reform into their original shape and function.

By now I'm sure you don't need me to point out how extraordinary these creatures are. I'm sure you also don't need me to ask what sort of sadist would kill and pulp someone, then pass their victim through a sieve, purely to satisfy his own curiosity as to whether this other would be able to re-form?

The next immortal creatures I want to talk about are *Turritopsis dohrnii*, called "immortal jellyfish" because, having grown to sexual maturity, they can, when exposed to environmental stresses or physical damage, or when sick or old, revert to the jellyfish equivalent of infancy. Then they can mature again, revert again, and so on, forever.

As evolutionary biologist and all around fan of coral and jellyfish Maria Pia Miglietta puts it, "instead of sure death, [*Turritopsis*] transforms all of its existing cells into a younger state." *National Geographic* describes the process: "The jellyfish turns itself into a bloblike cyst, which then develops into a polyp colony, essentially the first stage in jellyfish life. The jellyfish's cells are often completely transformed in the process. Muscle cells can become nerve cells or even sperm or eggs."[82]

And then they can mature again. And then revert. And then mature.

So, if you get mangled in a car wreck, or if your kidneys fail, or if you get Crohn's or leukemia, or if you get old and tired, no big deal. You just revert to infancy and grow up again.

Of course they can still be eaten by predators and so on. And also of course industrial humans have made a mess associated with these creatures. This species originated in the Mediterranean, but global shipping has caused it to spread around the world in what Miglietta calls a "silent invasion," an invasion that was only recently noticed. Gosh, we have a biologically immortal invasive species, spread by the global industrial economy, with for the longest time no one even noticing. What could possibly go wrong?

It gets worse. There is, according to James Carlton, a marine scientist at Williams College in Massachusetts, a "growing fleet" of unrecognized, invasive invertebrates.

Carlton also noted that this new discovery that the immortal jellyfish has overspread the globe highlights "our remarkable underestimation of the extent to which the ocean has been reorganized."[83]

"Reorganized." I guess that's one word for it. And I guess understatement sometimes has its virtues.

I'm not sure, however, if this is the language most of us would use were *we* the victims of this "reorganization." Then we might say "theft," "murder," "extirpation," "genocide," "ecocide."

Stolid scientists are saying that salt water fish could be extinct in thirty years. *Reorganized* is not the word I would use to describe what is being done to the oceans.

Note also his use of passive voice, instead of active voice, with the latter's subject/verb/object assignment of causal responsibility. Who is causing the oceans to be "reorganized"? The oceans aren't being "reorganized" by the actions of some random entity. They aren't being "reorganized" by the actions of the sun, or God, or Martians, or malevolent aliens, or whales. This culture of human supremacists is causing the oceans to be "reorganized," read "causing them to die."

So, members of this human supremacist culture are materially benefitting from actions—read, theft and murder—leading to the "reorganization" of the oceans. And, what do you know, it is members of this human supremacist culture who are consistently underestimating the extent of the harm caused by this theft and murder. Remarkable. Who'd a' thunk that thieves and murderers who consider themselves superior to those they steal from and murder, and indeed who perceive these others they are stealing from and murdering as not even having subjective existence, but rather being either resources to be exploited or competitors to be ruthlessly destroyed (or sometimes both at the same time), would underestimate the harm caused by the thefts and murders they perpetrate?

And now we come to lobsters. Lobsters can't claim any of the fancy forms of immortality: they don't regenerate if you cut them in two, they can't survive in outer space, they can't survive extremely cold temperatures. They sure as hell can't survive boiling water. They just don't get old. To reuse Caleb Finch's delightful phrase, they undergo "negligible senescence." If you recall, as our cells reproduce, the ends of the chromosomes can become frayed, leading to many of the problems of old age (although I'm not sure frayed chromosomes are responsible for *all* the problems of old age, like that young neighbor who just moved in and who blasts that lousy music that all sounds the same—boom, boom, boom—and why don't kids these days have any manners? When I was a kid . . .). There exists an enzyme called telomerase that repairs the telomeres that protect the ends of our chromosomes. Most vertebrates express this enzyme in the embryonic stage, when their cells have to reproduce quickly. But then most of us stop expressing this enzyme (which is a good thing, in general; expression of telomerase in damaged or old cells can lead to cancer), which means our cells become damaged as they reproduce. Which means we age. Lobsters and a few others continue to express telomerase throughout their lives. This means that lobsters don't slow down, weaken, or become less fertile as they get older. In fact, older lobsters may reproduce more easily than younger ones.

Perhaps lobsters say, "Viagra? We don't need no stinking Viagra."

The standard caveat applies: immortality, in this case, means they don't die of old age. They can still get sick. They can still die from exposure to pollution. They can still be eaten by predators, although by the time lobsters get to be pretty big, only humans, cod, or seals among megafauna can still generally kill them.

We could go through a list as long as the world, with words as tiny as nanobes or viruses, and still never come to the end of different ways that different beings are extraordinary. Many turtles resemble lobsters when it comes to aging. Autopsies reveal that the internal organs of many types of elder turtles have not degraded; inside, they're still teenagers. They died, but not of old age. Or there are pythons, whose hearts expand by 40 percent in the days after they eat a large meal; a human heart expanding similarly would be a precursor to death. Naked mole rats (who look scary but are evidently kind and gentle) are impervious to pain caused by acid.[84] And we all know that grizzly bears hibernate, but did we all know that when they're gorging in preparation, they may consume 50,000 calories and gain sixteen pounds *per day*? And did we know that when they hibernate for up to seven months, they don't eat, drink, urinate, or defecate? When they hibernate, they turn off their insulin receptors, becoming diabetic. Likewise, through their hibernation, these bears shut down the functioning of their kidneys. As a result, their kidneys scar and their blood toxifies. But come spring, their insulin response returns to normal, as does their kidney function. Their kidneys suffer no lasting damage. And to bring us back to forms of immortality, there is the Leach's storm petrel, a tiny bird who lives more than thirty years, and who is the only known animal whose telomeres grow longer with age. And of course there are those glass sea sponges I mentioned, whose lifespan can exceed ten thousand years.

And we haven't even talked about long-living coral (such as the black coral off Hawaii who is more than 4,000 years old), fungi (such as the honey mushroom in Michigan who is 1,500 to 10,000 year old) or plants (such as the Neptune grass in the Mediterranean who is 100,000 to 250,000 years old) . . .

* * *

My mother is getting old. Someday she will die. I don't know how I will survive that. I don't know how any of us survive the death of a beloved parent. And many of us—human and nonhuman—do. The death of a parent, and ultimately the death of this parent's child, too—that is, all of us—is one of the costs of entry to this wonderful thing called life through sexual reproduction.

And the knowledge that we all die is one of the primary causes of all monotheistic religions,[85] with their gates of various heavens standing open and waiting for their respective adherents to shuffle off this mortal coil, after which, if these adherents followed the rules of that particular religion while they were on earth, they will receive the blessing of immortality normally reserved for planarians, lobsters, bacteria, and others of the evidently spiritually pure.[86] A fear of death and a yearning for immortality is also a primary motivator of much human supremacist science, not only in its desire to assume the omnipotence, omniscience, and immortality of the Abrahmic God, but more prosaically, in its attempts to mine all of the immortal nonhumans I just mentioned to see how we can steal their immortality for our own. Rare indeed is the article about any immortal or even long-lived nonhuman that does not conclude with descriptions of scientific attempts to create human immortality.[87]

When my mother's grandmother was dying, my mother, fairly young at the time, comforted her by saying, "You'll just go to sleep, and when you wake up Grandpa will be there with you."

I do not know if when the time comes I can comfort my mother in this same way, telling her that she will sleep, then wake to find herself with her parents and others whom she loved and loves dearly. I do not know that I have that faith. I do not know what happens when we die. The lights may simply go out, and we sleep dreamlessly forever. Or we may dream, and as our bodies decompose our dreams may more and more resemble the dreams of the land, until ours and theirs are all the same. Or for all I know, human supremacists come back as their victims, and spend their short miserable lives unsuccessfully trying to communicate to their boneheaded former-comrades-in-superiority that mice or lizards or pigs or soy beans really do have subjective existence, before they are tortured to death. I don't know.

What I do know is that when my animal friends have died, they have always come to me in dreams three nights later, happy and playing. Perhaps this is what I will tell my mother, that she will close her eyes, and three nights later I will see her, and she will no longer be in pain, no longer be functionally blind, no longer have a broken neck and shattered forearm, and she will be with those, like her parents, whom she loved and loves.

❄ ❄ ❄

Lobsters, hydras, planarians, bacteria, and many others don't have to worry about all of this. They don't *have* to die.

As if all that weren't enough of a blow to the human supremacist ego, another massive blow can occur when first we recall that bacteria outnumber human cells ten to one in "our" body, and then we realize that many of these bacteria will outlive us.

Think about it.

If these bacteria had a supremacist mindset, they may have considered themselves to have merely hitched a ride in us, exploited our big, juicy bodies for as long as it gave them pleasure to do so, then when we died, consumed as much as they could before discarding what's left, just as they would discard the useless hulk of any other inferior being: with some of the most arrogant of the bacteria supremacists stridently objecting to the use of the word *being* to describe humans, because it's clearly a projection of bacteria features and being-hood onto those, like humans, ~~who~~ that do not subjectively exist and ~~who~~ that consistently fail every test of intelligence or self-awareness bacteria devise. These insignificant others—like humans—only exist, the bacteria supremacists assert, as mobile resources to be exploited.

❄ ❄ ❄

Nah, bacteria are too smart to believe something so stupid and community-destroying. And they've been around long enough for them not to have this worldview; a supremacist worldview is not sustainable, as it leads to the Cheat behavior described so eloquently and accidentally by Richard Dawkins. I'll be clear: mechanistic scientists love to yammer on about how evolution is based on traits that facilitate survival, and how traits that do not facilitate survival are selected out by causing the bearers of those traits to go extinct. Well, a supremacist mindset is maladaptive, in that it does not facilitate survival, for reasons that are becoming more clear by the moment. Likewise, a perception that natural selection is based on ruthless competition—which is closely allied to and interdependent with a supremacist mindset—is maladaptive. Both lead to behavior that destroys the landbases on whom those who hold these mindsets depend. And if you destroy your landbase, ultimately you destroy yourself; you cannot survive without the source of your life. The fact that I even have to say this is a measure of this culture's insanity.

While human supremacists are busy destroying life and asking why any being would "waste energy" helping "competitors," bacteria are busy helping each other gain resistance to antibiotics, helping each other to survive.

❀ ❀ ❀

Lately I've learned I have a heart condition, and that has me thinking about teleology. It also has me thinking about false solutions to environmental problems. Let's talk about the second one first.

The symptoms of heart problems sometimes resemble symptoms of heartburn, or acid reflux. A few days after I started feeling pain in my chest I got a resting electrocardiogram, and it was normal. I was going to get a stress test (where they stress your heart by making you walk faster and faster on a treadmill) but the testing machine at my doctor's office was broken. That's unfortunate, because there the test would have cost two hundred dollars, and the same test at the local hospital was over eleven hundred. I had no insurance. So I decided the problem must be heartburn. I know that's stupid logic, especially when a heart attack can kill you, but such is how people without health insurance sometimes make life-and-death decisions.

I'm generally a good patient, and the pain in my chest was severe enough to make me the best damn patient around. I took my heartburn medicine regularly, and made sure to be sitting up for two hours after eating.

The pain lessened. Or did it? Sometimes it was better, sometimes not. Yes, it was better. A lot better. Oh, wait, it still wasn't so great. Actually it was pretty bad.

I kept up with my regimen.

Six weeks later I became one of those people for whom Obamacare works as intended. I got great insurance. The next day I went for and completely failed a stress test, revealing heart disease.

My point is that it didn't matter how rigorously I followed my treatment regimen for acid reflux when the problem was actually my heart. We all see the larger lesson, right? If your diagnosis is wrong, it doesn't matter how carefully you follow the prescription. And this, of course, applies to the murder of the planet as well; any solutions to environmental problems following from a misdiagnosis won't, except I suppose by sheerest accident, lead to useful solutions. And since unquestioned assumptions *are* the real authorities of any culture, any solutions to environmental problems that emerge from a human supremacist mindset will, not surprisingly, serve human supremacism and not, once again except by sheerest accident, the real, physical world.

The heart disease also has me thinking about teleology, and how in this culture only human-created function counts as true function, which really means that only consciously-created function counts as true function, since our bodies aren't considered "human," but only animal matter; our minds are what's important.

Here's why I'm thinking about this. After the stress test, I went to see a cardiologist (and used my sparkly new insurance card; how cool is that to walk into a hospital, hand a receptionist a piece of plastic that is not a credit card, and get to have someone help you to be healthy?). He explained to me that we needed to do more tests, but preliminary indications were that one of the arteries into my heart is partially clogged. This means I'm getting insufficient blood flow to my heart, which deprives my heart of oxygen. This causes many problems, including the pain and breathlessness. Oh, did I forget to tell you I was breathless? I tried to forget that symptom as well, even as it was happening. . . . Not that we've ever seen anyone attempt to forget any symptoms on a larger, ecological scale . . .

A while later I asked him, why, if the problem was my heart, did taking the heartburn medicine seem to help? I went from debilitating chest pain as I was gasping for breath—and yes, I should have gone to the emergency room, but many of us don't always respond appropriately to emergencies and instead hope the crises go away on their own, not, once again, that we've ever seen anyone ignore crises on a larger scale either—all the way down to reasonably serious discomfort and a mere shortness of breath.

He said, "It wasn't the heartburn medicine. The improvement came because your body performed its own bypass surgery."

What did he just say?

"It sensed you weren't getting enough oxygen, so it grew new capillaries to go around the clog to supply oxygen to your heart."

I couldn't keep from exclaiming, "Just like plants know where to grow limbs and roots!"

I don't think he heard me. Or maybe he was just polite enough to pretend. In any case, he said, "It's not all good news, though, since sometimes these capillaries aren't stable or strong enough to not blow out. But they're certainly the body working to repair itself."

"Teleology!" I said.

Actually, no I didn't. This was my first meeting with this doctor, and I didn't want him to think I was crazy. If it would have been my third or fourth visit, no problem.

Think about it: how do the capillaries decide where to grow? Who makes those decisions?

Why is it that when humans consciously conduct heart bypass surgery, we perceive it as miraculous and a sign of our superiority as a species, but when our bodies do it without our minds' conscious intervention, we don't see this as a sign of superior intellect on the parts of our bodies?

And why did I just identify with my mind and not my body?

It all comes back to that belief that only human functionality is true functionality. And in this and many other cases, "human" doesn't include even our own bodies.

❀ ❀ ❀

I just read that sperm whale feces are central to the overall health of the South Pacific Ocean. This is because, as opposed to oceans where nitrates might be the limiting factor for the growth of plankton and so on, in the South Pacific the limiting factor is iron. And it ends up that sperm whales individually and collectively move massive amounts of iron from deep waters to the surface, where it becomes available to the plankton. How? By eating and then defecating. Sperm whales are known for being able to dive very deep in order to feed on squid, who are, evidently, iron rich. The whales swim back to the surface, then defecate, to the delight and health of just about everyone in the region.[88]

Whaling operations harmed the populations of sperm whales, and consequently, the health of the entire community.

This is what happens every time we forget that the natural world is full of intelligence, and knows far better than we do what it is doing.

❀ ❀ ❀

Today I had an engaging conversation with one of my neighbors, a genuinely good man who is the local Seventh-Day-Adventist pastor, and who has the wonderfully Dickensian name of Mason Philpot. I told him about my heart condition, and about my body throwing out these new capillaries.

He responded, "People often tell us that political and economic issues are too complex for us to understand, but I think mainly the problem is that those in power hide facts from us, and if we knew more, we would understand more, and we would be able to better respond to them."

I wasn't sure where he was going with this. I feared he was going to add his voice to the chorus of human supremacists who argue that if we just had more information about "how nature works" then we would be able to better manipulate the natural world.

I needn't have worried. He's a good and humble person. He went another direction entirely, and said, "But that's not what happens with our bodies. Honestly, the more I learn about our bodies, or about nature, the more I'm filled with awe at the beautiful, complex mystery of it all. It's all so powerfully and incomprehensibly beautiful."

Just wait till he hears about the role of sperm whale feces in the health of the South Pacific . . .

CHAPTER SEVEN

NARCISSISM

Narcissism falls along the axis of what psychologists call personality disorders, one of a group that includes antisocial, dependent, histrionic, avoidant and borderline personalities. But by most measures, narcissism is one of the worst, if only because the narcissists themselves are so clueless.

JEFFREY KLUGER

If I killed them, you know, they couldn't reject me as a man. It was more or less making a doll out of a human being . . . and carrying out my fantasies with a doll, a living human doll.

SERIAL SEX KILLER EDMUND KEMPER

It's not that I don't understand this whole notion of valuing what we create more than we value what nature creates. When I was a child—and of course now, I'm horrified and ashamed I did this—I loved making terrariums, even caught lizards and snakes to put in them. I dreamt and dreamt of making the perfect terrarium, which would be so large that none of those who lived in it would ever know they were in a terrarium, but still it would be my creation. Never mind that there was a wild world full of lizards and snakes and everyone else already outside my door, with no need for them to be deceived into thinking they weren't captives, since in all reality they weren't.

So I understand the impulse. What we create and control has value. What nature creates does not.

I still like planting seeds, and I pay closer attention to the seeds I plant than I do to the native seeds who sprout in this forest. Likewise, I love putting food scraps into the forest, and watching for when they're eaten, whether by big creatures, in which case the scraps simply disappear, or smaller organisms, in which case the scraps can take weeks or months to change color, collapse in on themselves, and finally become someone else. And the point is that I get more excited watching this process for the pumpkin scraps I place in the forest than I do for the dried berries hanging on the salal shrubs. The former are *my* contribution, and therefore special.

I understand all this. I also understand that this overvaluing of our own creations and creativity and undervaluing nature's creations and denying nature's creativity helps explain many things about this culture. It helps explain how an astronomer can say we need to explore Mars "to answer that most important question: are we all alone?" as this culture destroys life on this planet. It helps explain how so many foresters can continue to claim, as their "forestry" destroys forest after forest, that "forests need management." It helps explain how people keep trying to "manage fisheries" as they wipe out species after species. It helps explain how even so many so-called environmentalists state explicitly that they are trying to save, not the planet, but civilization, which so many perceive as humanity's—and thus the universe's—most important creation.

As opposed to perceiving life itself as the universe's most important creation. Or the universe itself as the universe's most important creation.

Imagine living in a culture sane enough to perceive life on this planet as more important to save (and worthy of saving) than this culture that is killing the planet. Imagine how quickly and dramatically our culture would change if sufficient numbers of people were sane enough to perceive this, and to act on this perception. Of course, if we lived in a culture that was sane enough to value life on the planet more than this culture, we wouldn't be living in a culture that's killing the planet.

This extreme valuing of what humans create and equal devaluing of what nonhumans create helps explain why scientists get so excited about "creating life" in a laboratory, as, once again, this culture destroys life on this planet. If we create an enzyme, that's worth far more than the world creating entire oceans full of life as varied and wondrous as goblin sharks, sea horses, angler fish, bull kelp, Portuguese man o' wars (who are not, in fact, single organisms, but communities of mutually dependent organisms) and the blue dragon sea slugs who eat them, then store their most venomous stingers to use for defense (and humans are the only ones who use tools?), and on and on.

Here's a question: who figured out penicillin's antibacterial qualities?

If you answer Alexander Fleming, I can, following *QI*, subtract points from your score and make a general hullabaloo. Then if you answer again with other scientists like Florey, Chain, or Heatley, I'll subtract even more points and make even more of a hullabaloo. At this point you might get frustrated, call me a pedant, and throw out the name of Ernest Duchesne, a French physician who noticed that Arab stable boys at an army hospital intentionally kept their saddles in dark, moist rooms to encourage the growth of mold on them. When asked why, they said the mold helped horses heal more quickly from saddle sores. Duchesne then did some studies, wrote them up, and sent them to the Pasteur Institute, which didn't even do him the courtesy of acknowledging receipt.[89] But here's the thing: if you throw out his name, I'll subtract even more points and make even more of a hullabaloo.

So then let's say you do some heavy soul searching, and at last you say, "Fine, I get it. I was being racist. The real ones who figured it out were the Arab stable boys."

I smile and say, "That's a very important realization, but you lose ten more points." Then I make even more hullabaloo.

Surprisingly enough, instead of strangling me, you get up and start pacing the room. Five minutes pass. Ten. Then you turn to me, an intense look in your eyes, and say, "I've heard that during the Crusades, soldiers on all sides put moldy bread on their wounds. They discovered it helped their wounds heal. You've got to give me those points back now."

Sadly, this leads to more lost points, and to more hullabaloo. Fortunately, it does not yet lead to you strangling me.

You say, "Oh, now I really get it! I was still being racist, and excluding Indigenous peoples. I'm sure that some of them figured it out."

"I'm sure you're right, but you lose ten more points."

I can see your fingertips quiver as you hold yourself back from forming your hands into claws.

So let's get to the point. Who discovered penicillin's antibiotic qualities? Why, fungi did, a very long time ago, when fungi of the genus *Pen-*

icillium were trying to figure out how to keep pesky bacteria from eating food—humans sometimes call this "spoiling food," and I'm sure these fungi and bacteria say the same about some of the things we do—before the fungi could get to it. After all, bacteria reproduce a lot faster than fungi. Well, perhaps asked the fungi, what if we just trim their numbers a bit? What if we invent some way to kill the bacteria who try to eat our food? Let them eat some other cantaloupe, not this one. And thus, not only did the fungi invent, but also discover, and indeed figure out, penicillin's antibacterial qualities. The same is true of nearly all classes of antibiotics: they were originally discovered and put in use by either fungi or other bacteria.

Perhaps this is when you move forward to strangle me. Not for pointing out the human supremacism inherent in the way this question is nearly always asked and answered, but for being so damned annoying about it.

※ ※ ※

Why do we as a culture refuse to acknowledge the creativity and subjectivity of nonhumans? Why do we insist on our own superiority? Why do human discoveries or inventions count, and nonhuman discoveries or inventions not count? (Of course, discoveries or inventions by Indigenous peoples don't generally count, either.)

Seventeen years ago I wrote that so often the perpetrators of atrocities "share a deeply unifying belief in their own separateness and superiority, and a tightly rationalized belief in the rightness of their actions. The perpetrators share a deep fear of interconnection and of the unpredictability of a life that may end in death tomorrow, or not for a hundred years, but one that will nonetheless end. The psychologist Erich Fromm changed Descartes' dictum from 'I think, therefore I am,' to 'I affect, therefore I am.' If Gilgamish can cut down a forest, if he can make a name for himself, he has affected the world around him. If Hitler can 'purify' the Aryan 'race,' if he can become the progenitor of a thousand-year Reich, he has, too. If my father can make my teenage sister wet her pants from fear and pain, or if he can make me take his penis against my skin—and more broadly if he can destroy our souls . . . you get the picture. Frederick

Weyerhaeuser (acting now through the unliving yet immortal corporate proxy that bears his name) deforested first the Midwest, then the Northwest, and now wants the world. Fearful of life, the perpetrators forget that one can affect another with love, by allowing another's life to unfold according to its own nature and desires and fate, and by giving to the other what it needs to unfold. One can affect another by merely being present and listening intently to that other. All of this is true whether we speak of forests, children, rocks, rivers, stars, and wolverines, or races, cultures, and communities of human beings."

That is as true now as when I wrote it.

And it should not surprise us that this culture values what it creates and does not value what others create. What else would we expect from a culture of cheats, a culture based on systematic theft and murder? How deluded must we all be to believe that we can steal from a forest, call it "management," and expect for the result to be any other than the death of that forest?

❀ ❀ ❀

How about if I steal your liver, and then see how long you live? Then your pancreas. Then your stomach.

Why are you complaining? I'm just managing your body.

❀ ❀ ❀

Imagine this: a pancreas decides that the rest of the body doesn't consist of other members of a larger community, but instead that the pancreas is the supreme organ, really the only organ who matters. All of the other organs, indeed, all the other cells of any sort, are taking up space and energy that could be better used by the pancreas. The pancreas starts to grow, needs to grow. The growth of the pancreas is natural. This is what all organs do: they grow and fight and compete, and the strongest, most fit organ survives. That's life. And who needs a fucking appendix anyway? And there's plenty of surplus intestine. And for crying out loud, there are two lungs and two kidneys. It's incredibly selfish of the lungs and kidneys to have redundancy when the pancreas needs those wasted resources. So, good, done deal. One lung and one kidney gone. Next it needs to displace some gray matter. Oh, you say that gray matter won't survive outside the brain pan? Too bad, so sad; the rule of nature is adapt or die. The gray matter can adapt to living somewhere else, or it can die. And frankly human supremacists have shown they don't really need their brains anyway, having already substituted ideology for perception, thought, and reasoned and/or sensible response to external conditions.

The pancreas takes over more and more of the physical body, and more and more of the body's energy. The rest of the body is completely mortified, and is also dying. But that's not really a big deal, since the size and growth of the pancreas are all that matter.

A few of the cells in the pancreas think it's a really bad idea to kill off the body that is the pancreas's only home. But the vast majority of the cells in the pancreas are pancreas supremacists, who either believe that God—who looks like a giant Pancreas—gave them the body into which the pancreas is supposed to go forth and multiply, and over which the pancreas is supposed to have dominion and show good stewardship; or they believe that evolution gave the pancreas the tools to make the rest of the body jump through hoops on command. Pancreas über alles.

※ ※ ※

Many environmentalists are infected with this same delusional supremacism, and this same delusional valuing of what they do over what nature does. I remember years ago I was doing a Skype presentation to an audience in the northeastern United States. At one point a simple living "activist" started doing what so many simple living "activists" do, which is to dismiss organized resistance and say the only thing that matters is one's personal carbon footprint. He said, "You write books, and that harms forests. And you fly to do talks—except this one—and that adds carbon to the air." I said the hope, of course, is that the books or talks might have a net benefit for the real world, that the value of at least slightly changing discourse ends up helping the real world more than the harm done by these (and any other actions) within the industrial economy. He then turned it into a pissing contest regarding who had the personally smaller carbon footprint. This was a contest he couldn't win, not only because he had four children—he called them his "four delightful little accidents," four being two above replacement level—and was therefore ignoring the fact that having a child is the single most environmentally expensive action any industrialized human can take; but because, as I told him, every spare dime I make goes into protecting forty acres of second growth redwood, who would have been cut without my protection. And then there are the tens of thousands of acres of old growth forest I've played a small role (as a part of various organizations) in protecting. Because of the carbon these forests are sequestering, when I go on tour I could hire Lear jets to fly me to the various cities and chauffeured Bentleys to drive me to the events, all the while consuming appetizers of caviar served on roasted hummingbird breasts, and my carbon footprint would still be negative.[90] He responded, and this is the whole point, "But how many of these trees did you plant yourself?"

Why do I need to plant the trees? Why is protecting a standing forest not as good as recreating a forest who has been destroyed? And why not let the forest plant trees on its own? It has been doing it infinitely longer, and is infinitely better at it. The forest knows far better than I do what trees it needs, and where, and when, and knows far better than I do what trees should die, and where, and when. The forest knows one has to take care of one's own.

❋ ❋ ❋

I'm not, of course, saying that forests can never use help from humans (or salmon, or beavers, or mosquitoes), just like I'm not saying that rivers can never use human help in removing dams. I'm simply saying two things. The first is that protecting living forests from being cut down is at least as important to forests as helping them to regenerate once they've been harmed (just as protecting living rivers from being dammed is at least as important to rivers as is removing dams from those already harmed). And the second is that it's irrelevant to me whether humans or forests plant the trees, just as it's irrelevant to me whether humans or rivers take out dams. The important thing is not and has never been human agency and control. The important thing is the health of the forest, and the health of the rivers.

❅ ❅ ❅

There is something far more difficult to bear—at least for those of us who are alive, and who are not sociopaths—than the death of one's parent. This is the murder of the planet.

It is a beautiful, though far too warm late spring day. I used to sit next to the pond by my house for hours at a time, nearly every early afternoon through the sunny season. Now it is too often far too painful. Today I saw one dragonfly and three or four damselflies. Even six or seven years ago I would have seen a hundred of each. I saw no newts. A decade ago I would have seen a half dozen, and I've heard reports from old timers of seeing hundreds. I saw no tadpoles. I recall years ago seeing scores of tadpoles scatter from the shallows as my shadow passed over them. I saw no caddisfly larvae, no dragonfly nymphs, no butterflies.

This is what it is like to be a living being at this point, a living being who recognizes that the world consists—for now, at least, even if not for much longer—of more than just humans.

❅ ❅ ❅

I'd love to be able to say that I first learned of *The Sorcerer's Apprentice* by reading Goethe's poem when I was a child, doing a little light reading in the original German, of course. But the truth is that like so many people of my generation, I first encountered this story in Walt Disney's *Fantasia*. If you recall, during one segment of the animated feature, Mickey Mouse is an apprentice to a sorcerer. The story begins with Mickey carrying buckets of water down some stairs in a castle, then trudging to pour the water into a large basin. When the sorcerer leaves, Mickey gets the bright idea to cast a spell he's seen the sorcerer use to animate a broomstick. It works! The broomstick grows legs! And it walks! On Mickey's command it grows arms and picks up the two empty buckets. He leads it up the stairs and out to a fountain. The broomstick fills the buckets, follows Mickey down the stairs and to the basin, then empties them. So far, so good! Mickey is making matter and energy jump through hoops on command! Isn't that the point of life? Isn't that why all of evolution has taken place, so humans—*homo sapiens sapiens*, or in this case *mickey sapiens*

sapiens—can make matter and energy jump through hoops on command (as a translation of Goethe's poem puts it: "You're a slave in any case, and today you will be mine!")? With the broomstick firmly set to the task, Mickey takes a nap. What could possibly go wrong? He dreams that he has been able to likewise get the stars and ocean and weather to do his bidding. Sound familiar? All the while the broomstick keeps bringing in water. Mickey awakens to a flooded room. He doesn't know how to make the broomstick stop (just like we have no idea how to get rid of or clean up or fix so many of the horrors we've unleashed upon the world: GMOs, invasives, plastics and other endocrine disruptors, nuclear waste, heavy metals and other pollutants, depleted and polluted aquifers, biodiversity crash, global warming, or neurotoxins; and just like on the larger scale we seem to have no idea how to get rid of agriculture, civilization, or industrialization). So he splinters it with an ax. Unfortunately for him, each splinter regenerates—think: hydra—and suddenly he has something like 279 broomsticks carrying buckets of water into the now completely inundated room. The water threatens to take him away, drown him. He could die because of his arrogance. But at long last the sorcerer returns. He cleans up the mess, and uses a broomstick to swat Mickey on the butt as Mickey leaves the room. The end.

The story is a pretty straightforward metaphor: if you meddle in that which is beyond your capacity to comprehend, and/or if you attempt to control that which is beyond your capacity to control, you run a pretty good chance of causing catastrophe. Sound familiar? You may free yourself from some drudgery in the meantime, and you may dream about controlling the heavens and the seas and the skies, but you'll come to failure. Sound even more familiar?

I realized even at eight, when I first saw this in a theater, that at least for me the metaphor would have worked still better with nature as the sorcerer, the one who understands how the world is, and not merely a more experienced human—named, I later learned, after Disney himself.[91] The latter implies that if we can just become better and more experienced slavemasters, we'll be able to competently run the show. But of course that's nonsense, and completely counterfactual; as humans have attempted to control more and more of the biosphere, more and more of the biosphere is being "reorganized," read: murdered. I also recognized even as a child that the metaphor has at least one more point of departure from our situation regarding the earth, which is that the horrors created by this culture and its arrogance are creating permanent harm, and can't be forced by some feat of magic to disappear with no harm done. Despite those two points of dissonance between the metaphor and our current situation, the metaphor has stuck with me all these years, helped inform my understanding of the world, helped inform my understanding of this culture's stupidity, arrogance, and destructiveness. It has informed my understanding of this culture's politics, religion, economics, philosophy, epistemology, and certainly its science. It also, sadly, misinformed an essay I tried to write for a high school German literature class about Goethe's poem (which, as a youth, I never did read in its original German. And here is a free hint for those planning, as I did, on using *Fantasia* to cheat on an essay for your version of my aforementioned German literature class: in the original poem, the apprentice doesn't have mouse ears, and the sorcerer doesn't swat him on the butt).

※ ※ ※

When I think of *The Sorcerer's Apprentice*, I think of the catastrophic consequences caused by members of this culture believing they're able to control the natural world. I think of the United States Forest Service managing forests to death. I think of the United States National Marine Fisheries Services managing oceans to death. I think of the United States Corps of Engineers managing rivers to death. I think of the entire human supremacist managerial ethos of this culture, stealing and murdering its way across the earth, pretending it can steal from and murder complex natural communities without fucking them up. I think of members of this culture creating, then releasing, genetically modified organisms into the world. I think of members of this culture changing the weather. I think of this culture releasing poisons into the world, bathing it in endocrine disruptors, covering it with neurotoxins. I think of this culture creating plastics. I think of mountaintop removal.

What could possibly go wrong?

I think of this culture killing off passenger pigeons, Eskimo curlews, bison, pronghorn antelope, pollinators, whales, cod, seals, prairie dogs, keystone species, mother trees, mother grasses, mountains, prairies, rivers, forests, oceans. What could possibly go wrong? I think of articles I'm seeing now about the creation of mechanical "bees" to replace the pollinators this culture is murdering.

What could possibly go wrong with any of this?

I think of an article I saw just moments ago entitled "What If Mosquitoes Were Annihilated?" The article began by calling mosquitoes "Little. Annoying. Killing Machines." It went downhill from there, mainly citing CEOs of two corporations dedicated to wiping out mosquitoes—well, actually the CEOs are dedicated to making money as they wipe out mosquitoes. The CEOs described how fabulous it would be to eradicate mosquitoes, if we could only overcome the technical challenges. I mean, we don't really want to drain *every* wetland and denude *every* forest, do we? The real solution, according to them, is for their corporations to genetically modify mosquitoes to render them sterile, then release them into the wild. Then we can eradicate mosquitoes! And gosh, what could possibly go wrong? Well, not much, evidently. Or at least that's what one of the CEOs reassures us. And how does he know that not much bad will happen? Because of the name, silly! Don't you know that names humans give nonhumans determine the nonhumans' roles and functions in the real world? And if humans determine the nonhumans have no roles, well, then, they have no roles! It's naturalistic philosophy, dude! And etymology! As the CEO says, "To be honest, there isn't much evidence that mosquitoes do much good. In fact the name, anopheles. Anopheles mosquitoes are the ones that [sic] spread malaria and the Greek origin of their name actually means, of little use."[92] There you have it. And just so you know that this CEO puts his (receipt of) money where his (lying) mouth is, his corporation has tried to release these genetically modified mosquitoes in Panama without doing any sort of risk assessment on what this could do to the local natural communities.[93]

When I think of *The Sorcerer's Apprentice*, I think of the utter insanity, the megalomania, the narcissism, the sociopathology, of believing that you know better than the real world what is good for the real world.

I think of this culture.

And of course I think of naturalistic [sic] philosophy. I think of the belief that there is no true intelligence in nature, that the only true intelligence and true purpose and true function come from the truly brilliant minds of humans, of *homo sapiens sapiens.*

When I think of *The Sorcerer's Apprentice* I am filled with sorrow and rage and disgust at the arrogance of members of this culture, at their stupidity and selfishness, at their smugness as they murder the planet that is our only home.

※ ※ ※

I have written more than twenty books trying to describe this culture's destructiveness, in the perhaps vain hope that by pointing to and articulating some of the underpinnings of this destructiveness I might help give other people who, too, perceive this culture for what it is, the courage to also name this culture's destructiveness, and then to individually and especially collectively act to stop this culture from murdering the planet.

But the truth is that even after more than twenty books, after millions of words, I have still come nowhere close to plumbing the depths of this culture's depravity, insanity, arrogance, or its urge to control and destroy, all masked in terms of its own moral and intellectual superiority. And I fear that the horrible disease of human supremacism is so infectious and so deeply held among so many members of this culture, that to the last this culture will continue trying to manage, trying to control, trying to steal, trying to murder, till the biosphere of the entire planet collapses. And even then the human supremacists will try to continue in their ways.

The disease is that strong.

※ ※ ※

Here is a beautiful thing. As you know, bacteria outnumber human cells in our bodies ten to one. So we need to gain bacteria somehow; they have to come from somewhere. How does this begin to happen? It used to be thought that the placenta is a sterile place, and that babies encounter their first bacteria in the birth canal, where they meet, among others, *Lactobacillus johnsonii,* who is a milk-digesting bacteria. Normally, not many of these bacteria live in the vagina, but during pregnancy the population of *Lactobacillus johnsonii* there greatly expands, so that during birth the child is literally covered with them. Some of these bacteria are ingested, and they give the child the ability to absorb the mother's milk.

That would be beautiful enough, but more recent research shows that the placenta probably isn't sterile, but that instead, as one article puts it, the "placenta harbours a unique ecosystem of bacteria which may have a surprising origin—the mother's mouth."[94] Evidently the bacteria "somehow" (to use the scientific term) make their way from her mouth through her bloodstream and then either into the baby's bloodstream or into the baby's mouth and then gut through amniotic fluid. But how does this happen? How do the bacteria make their way? What are the relationships between mother, bacteria, and child? Who is helping whom, and how?

❋ ❋ ❋

I'm sorry, do you want to tell me again that there is no true function in nature?

❋ ❋ ❋

Or let's talk about snot. Like most of us, I guess, I haven't, except during a bad cold, given snot very much thought. And when I have thought about it, I've always presumed snot was made up primarily of white blood cells who gave their lives to fight off whatever illness I happened to be enduring.

But then I had a flight home canceled out of San Francisco, and had to spend a night in a hotel with not much to do, so more out of boredom than anything else I decided to put the internet to its highest and most important social use—and no, I don't mean starting a flame war because someone somewhere on the internet has some ridiculous opinion that *must be corrected right now*—but instead looking up more or less random factoids. Yes, the internet as the ultimate source of bathroom reading material. Among the questions I typed into a search engine was, "What is snot?"

And boy, am I glad I did. I learned that snot is extraordinary stuff. And I got the added bonus of finding out why so many kids eat their own boogers. It all came from a column called "Dear Science" in the Seattle weekly newspaper *The Stranger*: "Snot is your body's best defense mechanism, a sticky moat of protection against invading bacteria, viruses, and fungi. When it comes to where your body is open to the outside world, snot (more properly, mucus) provides a barrier against these alien invaders. Mucus, chemically, is quite fascinating. Sugar chains are attached to a protein backbone in mucus cells [which helps me understand why some kids eat it: it's a little bit sweet! Kids: try this at home! Preferably when your parents have formal company!], with the contraption released out into the open. These glycoprotein molecules rapidly and aggressively suck up water until they are plump, slick, and slimy [Plump, slick, slimy, and sweet: Yum! Sounds like a slightly used gummy bear!]. To an invader, this is a nightmare to navigate: tangled chains of protein and sugar, with every nook and cranny crammed with water molecules. (Boogers are when these chains become ever more tangled, finally resulting in a rubbery ball of partially dried-out snot. Neat!) The body adds antimicrobial enzymes to this mix, which digest the invading organisms as they slowly attempt to chew through this barrier and reach the thin underlying lining of cells. As the outer layers of snot are eaten or rubbed away, new layers are forming underneath, creating a sort of treadmill of slime for invaders to run on. Hence, during an infection, our bodies tend to make more snot in an attempt to run the invaders out. Although the surplus of snot is not much fun when we're sick, it's better than the alternative. People with cystic fibrosis have a damaged chloride receptor, preventing them from properly filling their snot with water. Without the nice slick snot, people with the disease are subject to all sorts of terrible infections—particularly in their lungs. Snot turns colors as the defensive enzymes within ramp up to attack invaders. Many of the attacks involve charging up metal ions—turning them into nastily reactive bombs against the invaders. For example, green snot comes from iron-ramped-up white blood cells. The human mouth remains the champion of sepsis—containing the most bacteria per unit of any normally functioning body part by far, aside from perhaps the later stretches of the gut. This makes the mucus of the lungs all the more remarkable. Initially, the air entering the lungs is full of pathogens. As the air takes many twists and turns down to the delicate and vulnerable alveoli, one by one the pathogens get stuck in the sticky mucus lining the passages. By the time the air reaches the alveoli, it has been scrubbed; air at the very ends of the lung is sterile—free of bacteria. All together, the body makes about a liter of snot a day—probably a bit more in the average toddler."[95]

What did I learn during my enforced stay in San Francisco? I learned that snot is a brilliant and elegant—and evidently for some, tasty—way for bodies to fend off infection (and have you ever before read the words "snot," "elegant," and "tasty" in the same sentence?). It's not crucial to the brilliance and elegance of this solution whether snot evolved over billions of years of random mutations, or because bodies have intelligence, or nature itself has intelligence, or, for that matter, if some god or gods are behind it. The point is that it's a much better solution than *I* could have come up with, that's for damn sure. The real point is, do you still want to say that there is no true function in nature?

❊ ❊ ❊

How you perceive the world affects how you behave in the world. If you perceive only human constructs as having meaning or function, then you will overvalue human constructs. And if you perceive nonhumans and their creations as not having meaning or function, then you will undervalue nonhumans and their creations. The same is obviously true for those who perceive the creations of males or whites as being more important than those of women or people of color.

It is crucial for those who are destroying the planet to insist that non-humans have no inherent and true functionality, because if species do not serve true functions, the larger communities they are part of won't suffer when humans eradicate them. They seem to believe they can destroy the great schools of fish without harming oceans, clearcut forests without harming forests, dam rivers without harming rivers, and so on. Human supremacists are maintaining this belief even as they cause the planet to die. But that is never what is important to them; even life on earth matters less to them than their feeling of superiority. Life on earth doesn't matter to them except as it affects their ability to maintain this way of life.

❊ ❊ ❊

Having grown up in the arid western United States, I've thought a lot about water rights, and how these rights to water are allocated. Generally it is through something called "prior appropriation," also called the "Colorado Doctrine" after an 1872 Supreme Court ruling. In a nutshell, prior appropriation says that the first person (or economic entity) to use water from a river or other source for what is defined by this human supremacist culture as a "beneficial use" has the perpetual right to continue using that same amount of water for that same use. A phrase to describe it is, "First in time, first in right." Anyone who comes along later can use some (or all) of the remaining water for the same or some other "beneficial use" provided the new user doesn't impinge on the rights of those who came before. These rights then become property, and can be bought and sold like deeds or other markers of ownership. So let's say a mining corporation is going to use a lot of water for some planned operation. And let's also say that all the water rights to the river have already been claimed. The corporation couldn't use the water from the river till it bought the rights to do so from enough owners of already-allocated rights.

Why do I mention this? Because the definition of "beneficial use" ties right to that same old ridiculous naturalistic [sic] belief I've been hammering in this book: that the only true functionality is human functionality. "Beneficial uses" are generally defined as industrial, agricultural, and household uses. And the inclusion of "household uses" is a Trojan Horse, since more than 90 percent of all water used by "humans" is used for agriculture and industry, which means that "beneficial use" is for all practical purposes defined as industrial and (industrial) agricultural uses.

There goes the world.

Of course, any worldview that was not human supremacist, and that was not in thrall to industrialism and to a way of life that is killing the planet, would recognize that the first beings to have beneficially used water from rivers are the rivers themselves, and the fish who live in those rivers, and the forests who live with the rivers, and the oceans fed by those rivers, and so on. And the Indigenous humans who live by those rivers. Benefitting the real world, indeed benefitting anyone but members of this human supremacist culture, is not real benefit. It does not effectively exist.

But how do these supremacists believe the rivers became so fecund in the first place? It was through the beneficial use of the water by the rivers themselves, and by other members of their communities.

In the narcissistic worldview of the supremacists, the only benefits that really count are those accrued by the supremacists themselves. And so what the Colorado Doctrine means in practice is that the Colorado River no longer reaches the ocean. Nor does the Rio Grande. Nor do many other rivers the world over; this is what happens when you allocate 100 percent of a river to "beneficial uses": there is no water left for the real world. Likewise, this all means that the Columbia has been turned into a series of reservoirs, with disastrous consequences for all of those—human and nonhuman—who have the real original claims on the water, and who truly put the water to "beneficial use."

❊ ❊ ❊

The same doctrine applies not only to water. It is true of mineral deposits, where it's finders keepers, everyone-devastated-by-the-mine weepers. It's also basically the "doctrine of discovery," where any colonial power gets to rationalize taking possession of—that is, stealing—anything it claims to discover. In every case, discovery by nonhumans or by Indigenous humans doesn't count as discovery.

Of course it doesn't, because the only true functionality is industrial human functionality.

❊ ❊ ❊

Abusers often attempt to make their potential victims dependent upon them, so as to make these potential victims easier to exploit; a potential victim who is not dependent upon the abuser has more readily accessible choices, chances to get away. Even when the potential victim *does* have choices, it is crucial to the abuser to make it seem as though there are none.

First among those whom the abuser must convince of the rightness of his abuse and exploitation is the abuser. How can he sustain his abusive behavior over the long term if he does not believe his power is deserved and righteous and necessary and used for the common good? How can he feel all of these things if he does not perceive himself as superior to those he exploits? And how better to make himself feel superior to someone than to perceive this other as (and better, make this other) dependent upon him?

And how better to convince himself that those he exploits are dependent upon him than by convincing himself that he is the bearer of true meaning and true function; that the lives, actions, and achievements of those he exploits have no inherent meaning or function?

Which is how you end up with discourse as absurd as this culture's, with its talk of managing (read: killing) forests, managing (read: killing) oceans, managing (read: killing) wildlife, managing (read: killing) the entire planet.

Not only does how you perceive the world affect how you behave in the world, how you behave in the world further affects how you perceive the world. Enslaving, torturing, and killing the world not only proceeds from but also helps create a religion, a science, a philosophy, an epistemology, a literature, and so on—in short, a culture—that declares humans to be superior to all others and human function to be real function and human meaning to be real meaning.

It's a very bad cycle. And it's killing the real world.

CHAPTER EIGHT

REGRET

In looking back, I see nothing to regret and little to correct.

JOHN C. CALHOUN, WHITE SUPREMACIST AND
FIRE-EATING PRO-SLAVERY SENATOR

Rats experience regret when they make a wrong decision.

Researchers at the University of Minnesota noticed this, and so decided to "design an experiment to induce regret in rats and then measure behavioral and neurophysiological markers consistent with regret." The experiment was in itself benign, so let's leave aside the ethics of intentionally and systematically attempting to induce regret in another, and leave aside what we would call such a person were he to do so to another human being, especially one whom he was holding captive, over whose life he has complete control. Instead, let's for now accept this experiment's unquestioned assumptions.

The researchers differentiated disappointment from regret. Disappointment is a response to things not working out, they said, while "regret is the recognition that you made a mistake and if you had done something differently, things would have gone better." Keep this definition in mind.

The scientists trained rats to walk along a four-sided path. At each corner a short walkway led to a food station. Each food station had different flavored food, e.g, cherry- or chocolate- or banana-flavored pellets. Different rats, of course, had different food preferences. When a rat would reach a corner, a tone would sound indicating how long the rat might have to wait at that corner before receiving food. Rats would make reasoned decisions as to whether it would be worth their while to wait, for example, twenty seconds for a cherry-flavored pellet, or to move on to the next corner and hope the wait was shorter. The rats were generally willing to wait longer for food they liked more. All of this also, by the way, shows that rats have a sense of time; I know some people believe humans are the only creatures who have internal clocks, but that's the sort of counterfactual insistence on absolute species uniqueness we've come to expect from members of this culture. It's also important to note that the rats were able to decipher the time value of the different tones established by the humans. I wonder how often humans (including researchers) decipher messages rats establish for them. Or is this the same old human supremacist teleology, where humans are the only ones who create messages with meaning, while nonhumans at best react?

The researchers mixed up the wait times, so the rats wouldn't know at one corner what the wait would be at the next. They compared this to humans going to restaurants, not knowing until they got there what the wait would be for a table: "You can wait at the Chinese restaurant and eat there, or you can say, 'Forget it. This wait is too long,' and go to the Indian restaurant across the street."

The core of the experiment, according to science writer Mary Bates, was that "researchers wanted to know what would happen when a rat skipped a good deal and then found out the next restaurant was a bad deal. (In one example, a rat that [sic] had an 18-second threshold for both cherry and banana skipped the cherry option when the wait was only 8 seconds. Then it came to the banana option and the wait was 25 seconds.)

"In these situations, the rat stopped and looked back at the previous restaurant it had passed on. 'It looked like Homer Simpson going, "D'oh!"' says [researcher] Redish.

"Steiner and Redish compared the behavior of the rats in regret conditions (skipping a good deal only to find themselves with a worse deal) to what they did in disappointment conditions (they made the right choice—taking a good deal or skipping a bad deal—but the next restaurant was a bad deal, anyway). The rats showed three behaviors consistent with regret. First, the rats only looked backwards in the regret conditions, and not in the disappointment conditions. Second, they were more likely to take a bad deal if they had just passed up a good deal. And third, instead of taking their time eating and then grooming themselves after-

wards, the rats in the regret conditions wolfed down the food and immediately took off to the next restaurant."

The scientists also recorded neural activity, and found it was similar
to that in humans experiencing regret. Both journalist and scientist were
quick, however, to make sure we remember that there remains a chasm
between humans and nonhumans: "That doesn't mean regret is the same
in humans and rats; as Redish points out, deliberating over the choice of
flavored food pellet is not the same as deliberating over which college to
attend, and we don't see rats doing the latter."[96]

Of course we don't see rats doing the latter. Rats aren't given a choice
as to which college they will "attend," that is, in which cage they will be
imprisoned. They aren't given a choice as to whether the experiments they
participate in will be ones where they're given a choice of different flavored
pellets; or ones where they're intentionally traumatized, made to inhale lavender oil, then pickled alive before having their brains dissected to see if
the lavender oil helped reduce their anxiety; or perhaps ones where they're
put into jars of water to see how long they can swim before they give up
and drown; or experiments where rats are turned into alcoholics, and then
stressed to see if this makes them drink more (and by the way, many studies
include scientists addicting rats to various drugs; it ends up, however, that
whatever validity these studies may have extends only to *imprisoned* rats,
because wild and free rats aren't interested in becoming addicted, as they
presumably have better things to do; and what does *that* say about our way
of life?); or experiments where they're given hideous diseases or grievously
injured; or the $2.6 million 2009 study at New York University where
infant rats were given electric shocks while being overwhelmed with the
smell of peppermint (in the hopes that the infants would associate this
smell with their mothers, and so perceive their mothers and not the scientists as being their torturers), then after weeks of this and other torment
(which included stressing the mother so much that she in turn abused
the infants), they were put into pools of water with no way to get out so
the researchers could time how long the rats swam before giving up, then
pulled out of the water just before death so the scientists could implant
electrodes into their brains that released the active ingredient in hallucino-
genic mushrooms, then put back into the water until they once again gave
up, after which they were killed and dissected.

Of course rats don't regret what school they're killed at, nor do they regret never being able to touch the ground, nor that they'll never have natural interactions with their friends and relatives. They regret none of that, because they were never given those choices. Human supremacists don't care about choices made by free rats; we don't know if under natural conditions rats routinely regret roads (and cheeses) not taken. Their lives were under the complete control of someone else, with their "choice" reduced to which artificially-flavored pellet they may eat as a reward for performing tricks their owners lay out.

Now that I think about it, that's not so different from the choices made by many humans.

But there's a bigger question here than whether rats can feel regret over this or that wrong choice. The real question is, can humans feel regret? Do humans have the capacity to have that "recognition that you made a mistake and if you had done something differently, things would have gone better"? If they have that capacity, why the hell don't they manifest it? Oh, I don't mean over questions like whether to eat cherry- or chocolate-flavored pellets, or whether to attend this or that college. I mean real questions having to do with the real world. I don't see a lot of regret over the extermination of great auks, passenger pigeons, or Eskimo curlews. The only regret I see over the multiple decimation of the cod is that they're no longer there in numbers sufficient for profitable exploitation. I'm not seeing a lot of regret over actions that are leading to the murder—oh, I'm sorry, reorganization—of the oceans. What, by and large, has been the cultural response to the melting of the icecaps? Regret? Hell, no. The overriding response has been the money-lust of the few who will profit from the newly exploitable minerals and oil and from the much-anticipated Northwest Passage (typical headline: "Arctic Ice Melt Seen Freeing Way for South Korean Oil Hub"; typical quote: Korea "plans to add tanks for storing almost 60 million barrels of crude and refined products by 2020, about the same as Singapore's current capacity. The nation also seeks to leverage its energy infrastructure, which includes five refineries, to become Northeast Asia's oil hub, said Kim Jun Dong, the deputy minister of energy and resources policy."[97] And another typical headline: "Climate Change Tourism Comes to the Arctic: $20,000 Luxury Cruise to Sail the Once-Unnavigable Northwest Passage."[98])

And how much regret is the dominant culture manifesting over land theft and genocide against Indigenous humans? Certainly not enough to give back the land. Not enough to overturn a United States Supreme Court ruling that said, in essence, that if this way of life is based on land theft and genocide, then such land theft and genocide "becomes the law of the land, and cannot be questioned"[99] (and truer words have rarely been written by a Supreme Court Justice, nor have there been many more explicit articulations of the role of law in justifying atrocity and the legalization of both exploitation and unquestioned beliefs). Not regret enough for the United States to not have a national holiday named for the first European slaver to reach the Americas, on the anniversary of the beginning of the American Holocaust. Not enough to stop the ongoing land theft and genocide against Indigenous humans that is happening as you read this.

How you perceive the world affects how and what you feel. It affects whether and what you regret. So of course if your way of life is based on privatizing benefits and externalizing costs, and if you are raised to believe that it is not only acceptable but desirable and indeed natural and inevitable to exploit others, or that others were put here or are just here for you to exploit, you're probably not going to feel a lot of regret as atrocities are committed in your name against them. Exploitative behavior has become "the law of the land, and cannot be questioned." In other words, if you're a sociopath you're probably not going to have a lot of regret over harm you do to others, except insofar as it harms your ability to further exploit them.

Regret the extirpation of a species? Not on your life. Regret our not being able to exploit them further? Now we're talking.

This is one reason nearly all news articles about an endangered species must include reference to this species' financial value to the economy. From the perspective of human supremacists, financial value *is* value. The inherent value of the other—the value of this other to itself and to its family or community or larger biotic community—is either going to be ignored, or at best, grossly undervalued.

Only if there is no available substitute for the supremacist to exploit (until the substitute too is endangered, then disappeared) will the supremacist concern himself with this other's continued existence. And only in that case might the supremacist perceive this other's endangerment as at least a trifle worrisome.

To be clear, what supremacists regret is almost never the decisions they make that lead them to exploit and commit atrocities against others, nor the effects of their atrocities on everyone but themselves (and possibly members of their ingroup), but rather they regret only ways that prior exploitation has decreased their current capacity to exploit.

The scientists I just cited reasonably defined regret as the "recognition that you made a mistake and if you had done something differently, things would have gone better."

Let's talk about how some "things" have gone because of actions taken by this culture, and let's talk about whether this culture regrets these actions.

Let's talk about topsoil. It's gone. Around the world. The Levant. China. Southern Europe. Africa. Even places where this culture is a relative newcomer, like the American Midwest and the Canadian prairie, have lost up to 98 percent of their topsoil.

And how has this topsoil gone? Primarily through agriculture, and also through other forms of removing vegetation, like deforestation. For six thousand years agriculture and deforestation have been carving their way across the planet, and for six thousand years the planet has been losing topsoil.

Do humans regret these actions?

Not by any important measure. They certainly don't regret them enough to stop destroying topsoil. In fact, the murder of topsoil—you do know that soil is alive, and is the basis of all terrestrial life, right? Oh, but I forgot, only what humans do can have true function, so soil has no true function—is accelerating.

Forests, 98 percent gone. Do human supremacists regret murdering forests? Not enough to stop.

Prairies, 99 percent gone. Do human supremacists regret murdering prairies? Not enough to stop.

Wetlands, 99 percent gone. Do human supremacists regret murdering wetlands? Not enough to stop.

A stable climate. Gone. Do human supremacists regret destabilizing the climate? Not enough to stop.

Indigenous human cultures. Do human supremacists regret murdering Indigenous humans (for remember: Indigenous humans are, to human supremacists, below the human/nature divide in the Great Chain of Being)? Not enough to stop.

I could go on and on and on. Unfortunately, that is what human supremacists are doing when it comes to murdering the planet: going on and on and on.

Can human supremacists experience regret? I'm not seeing much evidence for that.

Maybe some rats should set up an experiment.

Nah. What's the point? When it comes to harming others, we know human supremacists can't experience regret.

❊ ❊ ❊

Of course human supremacists don't regret the ongoing murder of the planet, else they would stop it. Whatever regret they may eventually feel will only be because they wished they could have made a couple more bucks before the planet died. And of course they will regret that the wretched stupid weak goddamn Earth betrayed them by spitefully dying, which made it so humans could never reach our magnificent potential as the most glorious beings in existence, the ones whose brains are the most complex phenomenon in the known universe, the ones to whom every being should sing (to the tune of the famous doxology), "Praise Man From Whom All Meaning Flows, Praise Him all creatures here below, Praise Him ye heavenly host." Fucking ungrateful treacherous Earth robbed us of what we could have been, all by dying. And what the hell did we ever do to the earth to deserve this?

Apart from all that, human supremacists don't and won't regret the murder of the earth. Why should they? They still have televisions and computers and iPhones. These are clearly more important than a living planet. In fact, a living planet is only important insofar as it makes these other things possible. Televisions, computers, and iPhones have meaning. Zebras, musk oxen, water fleas, and sulfur shelf mushrooms do not.

The scientists defined regret as "the recognition that you made a mistake and if you had done something differently, things would have gone better."

So, based on the clearly articulated and lived values of human supremacists, there's no reason for human supremacists to regret anything that
has happened so far, because there's no reason to presume that they believe
they've made any major mistakes, or that they think they should have
done anything differently, or that they think that things could have gone
better. After all, they have their computers and iPhones. Never mind that
the oceans are being murdered. Who gives a shit, right?

And human supremacists call themselves intelligent?

I'm having trouble even granting them the phrase "aware of their surroundings."

CHAPTER NINE

THE SEAMLESSNESS OF SUPREMACISM

White [and I would say human] supremacy is the conscious or unconscious belief or the investment in the inherent superiority of some, while others are believed to be innately inferior. And it doesn't demand the individual participation of the singular bigot. It is a machine operating in perpetuity, because it doesn't demand that somebody be in place driving.

MICHAEL ERIC DYSON

Male [and I would say human] supremacy is fused into the language, so that every sentence both heralds and affirms it.

ANDREA DWORKIN

Let's turn how we perceive the world back onto ourselves. A friend of yours tells me that you're sentient, and that the two of you communicate. I don't believe your friend. She insists that you two have conversations all the time. I laugh at her. She asks what it would take for me to believe you are sentient and that you and she communicate. I, following what that mechanistic scientist said to me so many years ago, tell her that it would take her telling you to do something that is against your nature, and you doing it.

This is just the same old epistemology of control so well articulated by Richard Dawkins: we need to make these others jump through hoops on command. It's also the same old naturalistic [sic] notion, based on the Great Chain of Being, that what humans suggest has meaning, purpose, function, and what nonhumans do has none.

What gives me the right to propose those conditions? And what the hell are you supposed to do to prove me wrong? How should you go against your nature? Let's say suddenly a swarm of aphids begins crawling all over you. Can you let the nearby trees know they should prepare defenses? Can you persuade a horde of ladybugs to rescue you? What? You say the trees and ladybugs can't understand your muffled screams under the rising mound of aphids? Is that because the trees and ladybugs are too stupid to understand what you're saying? Or maybe it's because you can't go against your nature and speak their language using phero-mones. Maybe you're the stupid one. Maybe you're the one who can't communicate. I mean, for crying out loud, a little baby tree could do that! And you can't? What's wrong with you?[100] Or forget the aphids: maybe your friend can suggest you go against your nature by bashing your nose as hard as you can against a tree to make a small opening, and then using your tongue to pull a grub out of the hole. Or maybe you can dive naked into the Arctic Ocean and swim down to get some mussels to eat. If you can't act against your nature in any of these ways, you must not be intelligent, and I'm sure that you and your friend don't communicate at all.

If plants were to construct a Great Chain of Being, something I have no doubt they would neither be stupid nor arrogant enough to do, I could then see them suggesting that such human languages as English or Mandarin are rudimentary because they rely so heavily on the mechan-ical energy of vocalization and because they so willfully give such short shrift to other complex and deeply influential languages, such as those that use pheromones.

My point is that it is, once again, tautological to consider humans as somehow more intelligent or having lives more meaningful than others because our primary recognized languages use mechanical energy, as opposed to, for example, chemical energy. It's completely absurd. And lazy. And arrogant. And self-serving.

We can go through this same process for other arguments human supremacists use to rationalize their supremacism. But we all should be able to think our own way through these, right? Wouldn't that be some-thing members of the most intelligent species should be able to do?

So I'm only going to go through three more arguments.

❊ ❊ ❊

The first is that humans are superior because we're the only creatures who use tools. This is tautological. Here is a common definition of tool: "a device or implement, especially one held in the hand, used to carry out a particular function." Great! Humans, who have opposable thumbs, decide that a characteristic that defines them as superior is the use of tools, many of which require opposable thumbs. Yeah, and I'm superior because I wrote books and you didn't, remember?

Using a broader definition of tool as "an item or implement used for a specific purpose," humans, of course, aren't the only beings who use tools. But as we've seen—and this is obviously not unique to human suprema-cism—ideology nearly always trumps reality in how we make sense of the world around us. When we learn of nonhumans who use tools, we nearly always—just as with the discussion concerning the relationship between brain size and intelligence, or the discussion concerning plant commu-nication—change the discourse terms to make certain that we remain number one. It doesn't really matter whether we're talking about crows or fish or dolphins (who've been known to use tools to masturbate) or chimpanzees, or woodpeckers, or beavers, or bees building honeycombs, there's always a reason the tool usage isn't real tool usage. And it frankly all boils down to this: if humans do it, it's meaningful and real and shows true function; if nonhumans do it, it doesn't.

But there's another point to be made here, beyond this, and it has to do with sneezing, and with diarrhea.

If you've ever sneezed or had diarrhea, then you've probably been used as a tool, that is, as "an item or implement used for a specific purpose." You could and probably would argue that you're not an item or imple-ment, so by definition couldn't be a tool, but my argument is that nobody is an item or implement; trees are just as alive and just as much subjective beings as are you. You could also argue that people don't use trees as tools; they use wood, and just as someone could carve one of your bones into a tool, that's not the same as using *you* as a tool. I could then turn your argument around and say that it's one thing to use some *thing* as a tool, and quite another and more impressive task to get a subjective being to do your bidding, and even more so to get this being to like it.

And when you get right down to it, viruses, bacteria, amoebae, and so on basically use you as a sprinkler to broadcast their babies. When you sneeze, what are you doing? You're sending whomever caused you to sneeze out into the world, to find new food, I mean, tools, I mean, wonderful splendid unique human beings whom they can infect. And what is diarrhea? Remember how the bacteria can collectively decide it's time to make you poop? And to do so explosively, meaning you spread your feces (and the little darlings who caused it) all around, all the better to find new homes? And the thing is, in the case of sneezing, they get you to like it. In the case of diarrhea, probably not so much.

In my books *Songs of the Dead* and *Dreams*, I explored at length the question of how various parasites change the behavior of their hosts, and asked the question of who is really in charge in those situations. When a dog or skunk becomes rabid, he or she in some cases starts snapping at others while drooling; the parasite's new generation is in the saliva, and the parasite is causing the behavior changes, so the potential bites can infect another. I've already mentioned in this book the parasites who cause fish to swim to the surface and flash their bellies, making it easier for sea birds to ingest the fish, and thus the worms. Or there are liver flukes who move into the bodies of ants, take over their minds—or, for you human supremacists, their "minds"— leading the ants each night to climb to the top of a blade of grass and clamp down hard, then wait there till morning, in the hopes that overnight a cow or sheep will ingest the grass, and thus the ant, and thus the liver fluke, who then takes up residence in the ungulate. Or there are parasites who cause snails to climb to the tops of rocks and wave their shells around, drawing the attention of birds, who eat the snails, and thus the parasites.

And don't think humans are immune to this. We are, no matter how much we may pretend we are not, animals and parts of larger natural communities. Remember that the next time you sneeze or get diarrhea. Or the next time you scratch your butt; pinworms can live in our intestines, and breed near our rectums. In the early morning, the females will crawl out of the anus and lay eggs. The eggs make us itch. If someone uses his finger to scratch this itch, he gets the eggs on his finger. If, as some children do, he later puts his finger in his mouth, the eggs are home free. But even if he doesn't put his finger in his mouth, all is not lost. If he touches, say, an item of clothing or a piece of furniture, and someone else later touches this clothing or furniture, then later touches his or her mouth, the pinworms have found a new home.

Face it. They're using you. You're a tool. No, not that way. Literally.

And we can't pretend that the results of being used can be fatal for nonhumans, but not humans. Dysentery, caused by amoebae or bacteria, killed more people in the American Civil War than did bullets. Right now three-quarters of a million people die each year from dysentery.

❀ ❀ ❀

The point is that it's nonsense to say that humans are the only creatures who use tools, especially when humans are routinely used as incubators, sprinklers, means of conveyance, food, and so on by nonhumans.

❀ ❀ ❀

There's a sense in which it's already easy for modern humans to accept that we're used as tools. No, I'm not suggesting modern humans would accept we're being used as tools by mere nonhumans. Not at all. We could never accept that.

But most people accept that the capitalist system uses them as tools. And for decades now Richard Dawkins and others have been making the argument that humans are basically the tools of selfish genes. He says that these selfish genes "swarm in huge colonies, safe inside gigantic lumbering robots [by which he means you and me and everyone else], sealed off from the outside world, communicating with it by tortuous indirect routes, manipulating it by remote control. They are in you and in me; they created us, body and mind; and their preservation is the ultimate rationale for our existence. They have come a long way, those replicators. Now they go by the name of genes, and we are their survival machines."[101]

The selfish gene theory is extremely influential, with tremendous acceptance (acknowledged or not) within both the scientific community and the general (non-Christian) public. So on one level we're fairly used to being considered tools.

The point I'm interested in making here is not the one I've made so many times before, which is that the selfish gene theory so well serves the dominant culture by naturalizing oppressive and harmful and exploitative behavior. Some of the tag copy for Dawkins's *The Selfish Gene* lays it out plainly: "This book tells of the selfish gene. A world of savage competition, ruthless exploitation, and deceit." No wonder it's so popular; it's basically libertarianism and capitalism projected onto the natural world.

The point I'm more interested in at the moment is how much easier it seems to be for many people to accept Dawkins's language than it might be for many of us to accept that nonhumans use us, and more broadly, that nonhumans are capable of using tools.

It's very simple, I think. It has to do with what Mancuso said about so many people having trouble accepting that plants are intelligent, but these same people seeming perfectly willing to accept "artificial intelligence," because computers are our creations, and so reflect our own intelligence back at us. As opposed to plants, whose intelligence is other.

Reread the quote above by Dawkins. It is saturated with machine language. Hell, he calls *us* "machines." It is pure projection of machine onto the real world. And I think that is why people in this human supremacist culture so eagerly accept his logic (apart from the fact that it rationalizes exploitation and the destruction of community). Sure, I've got no problem saying we're controlled by machines. And in fact, I've got no problem saying those little bacterial gene machines give us big ol' lumbering robots the runs.

But what if we turn it around, say that communities of bacteria decide it's time to send some of their young into the great unknown in an explosive and exuberant spray, in some great diaspora, and they use your muscles, your body, to do it?

Suddenly, faced with the notion of an animate universe, the barriers go up: No fucking way; you've got to be shittin' me, man.

❊ ❊ ❊

The second human supremacist argument is the one that really blows all of my arguments out of the water. Instead of something like brain size, you could say, why don't we go straight to one of the central definitions of intelligence and talk about an ability to solve problems? That's pretty much checkmate, isn't it?

I agree. I don't think any intelligent person would expect a penguin to be able to adjust a carburetor. Of course, no intelligent person would expect me to be able to adjust one, either. But the fact remains that humans excel at solving problems. Let's ask that penguin to solve analytical reasoning problems of the sort we ourselves might try unsuccessfully to solve on the Graduate Record Exams and see how it does . . .

Question 1 (and these are real questions from a practice exam website): "A person starts walking north, stops after 15 km, then turns 45 degrees right. Next, this person turns 135 degrees in an anti-clockwise direction. How far is this person from the starting point and in which direction is this person facing?"

A penguin wouldn't be able to . . . oh, wait, actually a penguin probably could answer this question, and could also answer it while swimming in varying ocean currents, catching fish or krill, and avoiding leopard seals.

Question 2: "I promised my mother that I would meet her in the month of October, but it should be a Sunday. Thursday morning my mother called me up and reminded me to meet her by stating that 'Today is the 21st of August and September is coming very soon.' On which of the following dates could I meet my mother?"

So you're right: a penguin probably couldn't answer that one, although I'm guessing a lot of penguins see their relatives a lot more often than this person seems to.

But I get your point. Humans solve problems. That's how we got to where we are today, by solving problems. And I'm sure you know what I mean by "where we are today": on a world being murdered. And I'm sure you know that unfortunately, most people don't see it that way, when they see it at all, which is a problem these human supremacists can't seem to solve. Humans are great because we can solve problems that couldn't have been solved (read: created) by penguins, whales, octopi, bald eagles, tallgrass prairies, estuaries, mountains, or moonlight. You can't argue with this.

But I'm going to anyway. I have at least three problems with it. The first is that the notion that humans are superior because we solve problems is once again and as always tautological; humans define themselves in part by their ability to solve problems, or rather, some problems of a certain sort, solved only in certain ways; humans define intelligence and superiority as based in great measure on their ability to solve these particular problems in these particular ways; humans have therefore defined humans as having intelligence and being superior. Isn't it wonderful how it always seems to work out that way?

But it's even worse than this, because as I alluded to, it's all based on very specific and limited definitions of problems and solutions.

When I think about people who conflate intelligence with solving more or less meaningless problems in logic, the first person I think of is Marilyn vos Savant, whose website states she was listed for five years in the *Guinness Book of World Records* as having the highest IQ. The most well-known way she uses her intelligence is in a weekly column in *Parade Magazine* where people ask her brain teasers kind of like the ones above.

My main problem with her is not that sometimes when answering her questions she throws in her fundamentally authoritarian, misogynist, and pro-capitalist politics. My main problem is that for many years I've wanted to send her a couple of questions of my own, formed into sort of a story problem: "Decades ago a friend said to me that if the universe gives you a gift, and you do not use this gift in the service of your community, in the service of life, then you are not worth shit. I immediately saw the truth in what he said, and changed my actions accordingly. So, if you really have the world's highest (human) IQ, and if having the world's highest (human) IQ really means you are extremely intelligent, why are you using this gift to play parlor games, especially those where the real point seems to be for you or others to show off? The world is being murdered, oceans are being murdered, rivers are being murdered, mountains are being murdered, Indigenous human cultures are being murdered, hundreds of millions of women living right now have been raped, industrial civilization is causing the greatest mass extinction in the history of the planet, and this is how you choose to use your gift? If you really have been given this extraordinary gift by the universe, explain to whatever humans may be living in the wreckage of the world fifty years from now why they should not hate you for misusing and abusing your gift in this way, for completely failing to use your gift in the service of life."

I'm guessing she wouldn't print her answer in *Parade Magazine*.

Now, you could say she actually *is* using her intelligence to solve a problem, which is, how do you earn a living in this wretched capitalist system? *Parade* has a circulation of more than thirty million and a readership of more than seventy million, so I'm guessing she makes a decent living from solving a couple of absurd questions each week. I'm also guessing she could do this work in her bathrobe. Anyone in this wretched capitalist system who can make a decent living while wearing a bathrobe has certainly solved one problem.

But like many of the other problems she solves, there's no good reason to solve this one. She's married to Robert Jarvik, of Jarvik artificial heart fame, and was herself chief financial officer of Jarvik Heart, Inc. She doesn't need to do this. She clearly has her expenses covered and could, if she chose, use her intelligence for life-affirming purposes.

Sadly, I think vos Savant is not unique in using her gift in the service of trivia.

In my thirties I went to a Mensa meeting. I had no social life, and so with visions of literary salons dancing in my mind, I headed off to a bar where I'd read in the newspaper that they held their monthly meetings. Now, those of you who know my work know I'm a teetotaler, and that I don't really like alcohol culture. But I've been to a fair number of bars for blues shows, and this once for a Mensa meeting. The only people in the room were the bartender and the Mensa members. I sat down with the other eggheads, expecting a dazzling conversation about what we are personally and collectively going to do about the mess this culture has created. But that's not what happened. Instead the conversation mainly consisted of how various members of the group had found grammatical errors in the local newspaper and duly reported these to the offending reporters. One had even found a misused word in a headline! From there the evening basically turned into a free-form version of Trivial Pursuit where each person attempted to top all others in his or her capacity to share obscure mathematical, grammatical, historical, or mass media factoids. I lasted about a half an hour before I moved to sit at the bar and nurse a glass of water while I talked with the bartender about how she wished she could join a union.

I'd been expecting Lewis Mumford and instead got Marilyn vos Savant. In retrospect I'm not sure why I'd been fantasizing about literary salons; when I was in graduate school for creative writing, I went to a few parties—okay, one party; I said I had no social life—and it was more or less like the Mensa event, except that the trivia was centered around, for some reason I'll probably never understand, *The Brady Bunch*. Instead of talking to a bartender, I took a long walk.

So maybe the biggest sign of our intelligence is an attendance to trivia as the world burns.

How droll.

I'm still dancing around the subject, which is that humans are smarter—the best!—because we're so good at solving problems. Humans eradicated smallpox, once it had outlived its usefulness at killing Indigenous peoples. Humans landed on (and bombed) the moon. Humans created the Internet, which has as its primary economic use pornography, i.e., the objectification and degradation of women.

Wait. That doesn't sound so good. Let's try this again. So I'm writing this lying in a warm bed in a warm house looking out the window at a wind- and rainstorm. These wind- and rainstorms used to mainly come in December, but now, because of global warming these past few years, they mainly happen in March. I'm eating cashews from a plastic bag by the bed.

How many problems had to be solved in order to allow me to be here, out of the cold, first handwriting this and then typing it into a laptop computer? Well, there's the problem of converting a living forest to 2x4s for the house, the problem of converting a living beach to glass for the window, a living river to hydroelectricity to run the heater. There's the problem of clearing land for cashew plantations, and the problems of drilling for and then refining oil to use to transport these cashews from wherever the hell they were grown to my local supermarket, and the problem of creating plastic for the bag. There's the problem of inventing computers, and the problems of how to perform the toxic and water-intensive processes of manufacturing them. There's the problem of destroying local economies to force people off their lands and into the sweatshops that produce the computers. There's the problem of having (and paying for) large militaries the world over to enforce the dispossession of those who lived on the land where cashews are now grown, and to enforce the destruction of the local economies to force people into sweatshops. And on and on. And these problems were all evidently solved, because here I am, lying in bed typing on the laptop and eating cashews.

I'm sorry. I guess I'm still not behaving. I should discuss our intelligence, as evidenced by our ability to solve problems, with more reverence for our superiority.

I'll try.

Seriously, it takes a lot of intelligence to invent automobiles, doesn't it? Never mind that they are one of the three or four most destructive objects ever invented.

I'm still not getting into the proper spirit of human supremacism. Let me try again. We're the smartest species since we solved the problem of too many damn salmon by putting in dams. Er, no, that's not right. We solved the problem of how to violate—I mean harness for power—a river by putting in a dam that will kill the river, kill the now-submerged riparian zones, leach mercury out of the soil, and rob the area below of sediment.

The point I'm trying to make is that so many of the "solutions" to what members of this culture so often perceive as problems quite often lead to other problems, many of which probably could have been predicted were the people looking at these original "problems" both a) intelligent; and b) not evil. The only way I can see that someone putting in a dam could not predict that this would wipe out anadromous fish species, destroy the submerged lands, dispossess the human and nonhuman inhabitants, harm wetlands downriver, and so on, would be if those suggesting these "solutions"—e.g., dams—to "problems"—wanting electricity—were either a) unforgivably stupid; or b) unforgivably evil. Take your pick. None of this is cognitively challenging in the least. And of course you could say that they are simply unforgivably entitled, and don't care about those they harm, but a) that doesn't seem very smart; and b) that seems pretty evil. And we can talk all we want about claims to virtue and about social systems that reward atrocious behavior and so on, but beneath it all, this is what it comes down to: are they really that stupid, or are they really that evil?

You could ask, "Well, how else are they supposed to generate the electricity they need to run their factories?" But that's not my problem, and coming up with a "solution" to that "problem" is not the responsibility of those who will be murdered by the dam. If you can't generate electricity without causing significant harm to those humans and nonhumans who will not be receiving the benefits of this electricity, then you really are only "solving" the "problem" by foisting its harmful effects onto others, in which case you shouldn't be calling yourself superior or intelligent, but instead a thief of these others' lives. Theft and murder do not by themselves qualify you as intelligent or superior.

Or let's talk about pesticides. Recently scientists have discovered (for the umpteenth time) that pesticides are causing terrible ecological problems, killing off domesticated and native pollinators, other insects, birds, amphibians, streams, meadows, human beings.

I'm sorry, why is this surprising? Which part of neurotoxin did these people not understand? Was it the neuro, or the toxin? At my public talks I've long rhetorically asked, "Who's the idiot who came up with the idea of putting toxins on our own food?" But I think the real question is, "Who's the idiot who came up with the idea of bathing the entire world in neurotoxins?" That's fucking nuts. It's stupid. It's evil. It's both.

Shit. I'm still not doing a good job. I'm trying, though. Really, I am. Each time I start off by thinking about how great we humans are because we've solved this or that problem, but then each time I have to go and let out an explosive fart in the midst of our "We are Number One" Celebration Party by remembering and pointing out the negative consequences of so many of the "solutions" we come up with to "problems."

So I'm going to try again. Humans are clearly superior because we have invented plows (never mind for now that plows are probably *the* single most destructive human invention ever, and that agriculture was the single biggest—and least intelligent—mistake any creature has ever made; oh wait, I'm still not in the spirit of it), space heaters, locks, screws, levers, torque wrenches, artificial sexual lubricant (scores of different formulations, including wild cherry–flavored), laptop computers, telephones, airplanes (no prairie dog ever did that!), nuclear reactors, reading glasses, artificial hips, sleeping pills, refrigerators, reclining chairs, cameras, municipal water treatment plants, rockets to go to the moon, satellite television, the printing press, backhoes, and chainsaws.

Sure, all of these were in some ways solutions to perceived problems, but as a sign of greater sentience than others, or as some other sign of superiority, we run into the same old familiar problems.

The first is that it's once again and as always tautological: (some) humans invented refrigerators, therefore (some) humans decide that the invention of refrigerators is a relative sign of intelligence or superiority, thus (some) humans have determined that humans—especially the ones who invented refrigerators—are smarter and superior.

Aardvarks might choose other criteria.

※ ※ ※

The second problem is that the whole notion that only humans solve problems is ridiculous anyway. Don't you think human use of antibiotics created a problem for bacteria? And what has been the response of bacteria to this problem? Don't you think that bacteria creating resistance to these antibiotics and communicating this resistance to other members of their community is a solution to this problem? And what about the development of antibiotics in the first place by fungi? Wasn't that a solution to a problem?

If you just look around, you'll see a more or less infinite number of elegant solutions created by nonhumans and nonhuman communities to whatever problems these nonhumans and their larger communities have faced. It's a beautiful, wonderful process called life, and unless you have been rendered completely insensate by a grotesque human supremacist ideology, it's pretty fucking obvious.

Sadly, the one problem it seems nonhumans and their communities have yet to be able to solve is the sociopathy of human supremacists.

※ ※ ※

The third reason it's ridiculous to say that inventing refrigerators is a sign of intelligence and superiority is that if it is, what does that say about those Indigenous human cultures who never invented refrigerators (or cameras, telephones, or perhaps more to the point, iron blades, war chariots, galley ships, steel breastplates, tall ships, muzzle-loaders, breech-loaders, long-range artillery, machine guns, tanks, bombers, aircraft carriers, nuclear attack submarines, predator drones, and so on)? Does this mean they were less intelligent, because they didn't invent backhoes and chainsaws? Does this mean they were inferior? Are those really arguments you want to make? If so, are you really that racist? Because the belief that the invention of any of these "solutions" is a sign of intelligence and/or superiority implies that the failure to invent any of these "solutions" is a sign of a lack of intelligence and/or superiority, which means that it implies that those who have invented these "solutions" are more intelligent and/or superior to those who did not. This means the civilized are superior to and/or more intelligent than Indigenous peoples. Another way to put this is that they are higher on the Great Chain of Being than are Primitives.

I don't believe Indigenous peoples are less intelligent than the civilized, which means that the invention of refrigerators can't by itself be a sign of intelligence. I believe the Tolowa, for example, never invented chainsaws, backhoes, or refrigerators at least in part because they had such a different social reward system and such a different way of perceiving and of living in the world, that many of the problems that led to these solutions may not even have been perceived as problems. If you've not exceeded your local carrying capacity, and you rely on salmon for food, and you ceremonially smoke them, and if you recognize that your life is tied up in theirs, and if the salmon stay as common (and delicious) as they have been forever (as they should if you don't exceed local carrying capacity, either through overconsumption or overproduction or overpopulation), there's really no reason to invent refrigeration. The meat stays freshest in the river. And if you're not planning on conquering your neighbor, there's really no reason for you to invent chariots or steel breastplates or machine guns, is there?

CHAPTER TEN

AUTHORITARIAN TECHNICS

We become what we behold. We shape our tools and then our tools shape us.

MARSHALL MCLUHAN

There's another point I want to make here, one that was made best by Lewis Mumford. This is that technologies—and by extension, I would say many other forms of "solutions" to other forms of "problems"—do not exist in a vacuum. Technologies emerge from and then give rise to certain social forms. Mumford called the technologies and their associated social forms "technics." Technics, he said, can be democratic or they can be authoritarian. Democratic technics are those that emerge from and reinforce democratic or egalitarian social structures, whereas authoritarian technics are those that emerge from and reinforce authoritarian social structures. The distinction he made is both brilliant and simple: does the technology require a large-scale hierarchical structure? Does it reinforce this structure? Does it lend itself to the monopolization of the technology, and therefore to control of those who fabricate the technology over those who use it?

To put it in its simplest terms, is this technology something that anyone can make? Or is it a technology that requires massive hierarchical (and distant) organizations?

We can ask all of these same questions not just about technologies, but about all "problems" and "solutions." Authoritarian and egalitarian societies may look at the same situation and perceive entirely different "problems," to which they will perceive entirely different "solutions." These "solutions" will then lead to the societies becoming more or less authoritarian or egalitarian. We can also say that unsustainable and sustainable societies may look at the same situation and see entirely different "prob-lems" to which they will find entirely different "solutions." And human supremacist cultures and non human-supremacist cultures may also per-ceive different "problems" to which they will find different "solutions."

An authoritarian, unsustainable, human supremacist culture may look at a river and see both problem and solution. The problem: How do we power our factories? Solution: Dam the river for hydropower. An egali-tarian, sustainable, non human supremacist culture may look at the same river and see a different problem and solution. Problem: how do we live in place for the next twelve thousand years (the Tolowa Indians have lived where I live now for at least 12,500 years)? Solution: fold yourself into long-term interspecific communities such that these communities are healthier on their own terms because of your presence. Which means to respect and revere the nonhuman communities who share and are a part of your home, as you are a part of theirs.

Same river. Same original species composition. Different cultures. Different imperatives. Different attitudes toward the river. Different per-ceived problems. Different perceived solutions. Different results.

When authoritarian, unsustainable, human supremacist cultures encounter cultures which are none of these, they quite often conquer or destroy them. This is not only because unsustainable cultures must expand or collapse, but also because supremacist cultures by definition disrespect difference. But even when unsustainable cultures don't outright conquer or destroy those who are sustainable, the sustainable cultures may still find themselves harmed, or if you prefer, infected. For example, a sustain-able non human supremacist culture may face the problem of keeping warm in the winter and may choose as a partial solution the wearing of skins of fur-bearing creatures they have killed. After being contacted/infected by an unsustainable and human supremacist culture, they may begin to see their landbase differently. Now they may see the same forest, the same creatures as before, but the new problem is not, "How do we keep warm?" but rather "How do we make money?" or "How do we gain trade goods? How can we get some of those steel pots and steel knives, which are ever-so-much more useful than our clay or reed pots and our stone knives?" Their solution can then become, "By killing fur-bearing creatures to sell their pelts." And the culture has begun to move away from sustainability and interspecific cooperation and towards unsustainability and human supremacism. This is something that happened time and again across North America, as creature after creature who had lived with the Indigenous humans for millennia were quickly decimated, and the human cultures changed. Thus did technologies such as steel pots and steel knives play a role in changing cultures and landbases.

Let's explore some more examples of democratic and authoritarian technics. Baskets made from reeds would be a democratic technic, because anyone can make them. Obviously, some people can make better baskets than can others, and some people can learn techniques for making baskets which they can choose to share or not share with those around them. But unless you live in an area where there's only one small patch of reeds who could be claimed and guarded by someone trying to gain a monopoly on basketmaking materials (and even then, you could make them out of bark or grass or some other material), no one can physically control whether you do or don't make baskets. On the other hand, automobiles are a non-democratic technic. I can't build one from scratch all by myself. Automobiles require mines (which require forced labor of one kind or another) and mining infrastructures, they require transportation infrastructures, they require manufacturing infrastructures, they require energy infrastructures, they require infrastructures on which to drive your completed non-hand-made automobiles, they require crews to maintain all of these infrastructures, they require military forces to steal, I mean, conquer, I mean, protect and defend, the land where the mines are located, they require police forces to defend these infrastructures from those who unaccountably don't want these infrastructures on or near their homes, they require managers to keep the whole thing running, and autocrats of one sort or another to tell the military, police, and managers what to do.

Here's another example. Bows and arrows are a democratic technic. Anyone can make them (albeit poorly at first; I'm not saying there aren't skills to be learned, and I'm not saying that one person may not be more proficient than another; I'm talking about the capacity to construct and use a piece of technology free of distant control). Can you find materials to make a bow? Can you find materials to make a string? Can you find materials to make an arrow? Unless someone has a monopoly on feathers, you can even fletch it. And if you lose your arrows, you can make more.

Let's contrast that with guns. Immediately we again run into the problem of mining and smelting the metals. Even if you already have a gun, you still have to get bullets and gunpowder. You (and your community) are not autonomous, but can be controlled by those who have access to the raw materials and infrastructure to create the tools (in this case gun, bullets and/or gunpowder).

Let's do another. Passive solar is a democratic technic. Anyone can align a home to face the sun. Anyone can collect rocks to store the heat. No one controls the sun (and I can just see the look on the face of a capitalist as he reads this, then jots in his journal: "Note to self: find way to privatize the sun, claim ownership of it, then find way to force people to pay a royalty for each ray of sunshine they absorb. Should be no problem; I'll pay Congress to pass a law declaring I own the sun and then get the police to enforce it. Get lobbyists on this tomorrow.").

In contrast, solar photovoltaics, no matter how groovy and "alternative" they may seem, still require all of the infrastructures we mentioned above. They require an authoritarian social structure, with all that implies. They are in no way democratic or egalitarian, and in fact they aren't even particularly groovy. And they're incredibly environmentally destructive; take a look at photos of a rare earths mine.

The fact that anyone can make a piece of technology is not sufficient for that particular technics to be democratic. A small wooden plow, for example, would seem part of a democratic technic since anyone can make one, and pull it using his or her own strength. But members of a community being able to make a piece of technology is merely a necessary but not sufficient part of what defines something as a democratic technics. We must never forget that technologies affect our societies, and we must never forget to ask ourselves *how* these technologies affect our societies. Societies interested in sustainability and self-reliance have *always* asked themselves how new technologies will affect their communities. To not do so is a fatal mistake.

There are a few reasons we can say that plows underlie an authoritarian technics. The first is that to pull a plow is about as hard as to work in a mine, so plows lend themselves to the capture and use of slaves about as much as do mines. By 1800, about three-quarters of the people living in agricultural societies were living in some form of slavery, indenture, or serfdom, almost all of which could be blamed directly on agriculture. The only reason that isn't true today is that human slave energy has been temporarily replaced by fossil fuels; when these run out the human slave percentage will return to its former heights. And of course none of this is to speak of the nonhuman slavery upon which agriculture is completely reliant.

Another reason a plow-centered technics is authoritarian is that the product of the plow's use is food; if slaves are used to grow food for their owners, this means owners control the food supply. Controlling food supply is of course central to authoritarian regimes. The more necessary some product is, the more that control of the product by authorities leads to control of those who need the product; if those in power control my access to Cheez Wiz, they're not really going to gain a lot of control over me, but if they control real food, they control me.

The authoritarian nature of plows gets worse, though, precisely because of what a plow is designed to do: kill the native life in the soil. As Lierre Keith notes, "Agriculture is biotic cleansing: you take a piece of land, you clear every living being off of it, and I mean down to the bacteria, and then you plant it to human use. So it's biotic cleansing. The plow is a tool—really, *the* tool—by which this is done."

This means plows are part of an authoritarian technics. If biotic cleansing and conversion of a prairie, say, or a forest, to exclusively human use doesn't constitute the repression by one class of all other classes, I don't know what does. Just as deforestation harms those humans and nonhumans who live in and rely on the forests to be destroyed, so, too, destruction of prairies, wetlands, rivers (and oceans) and so on by agriculture destroys the lives and ways of life of the humans and nonhumans who live there. Human supremacists may not care, but then again white supremacists don't care about the effects of white exploitation of other races, and male supremacists don't care about the effects of male exploitation of women, except, in these cases, where it affects their own entitlement. Human supremacists are the same. And it's the same imperative.

And now we get to perhaps the most authoritarian part of this whole wretched technics: when a culture destroys its own landbase (through agriculture, through associated urbanization, or through any other means for any reason), it then has two choices: collapse, or take someone else's landbase. Since cultures rarely choose to collapse, this means once a culture has committed itself to an agricultural way of life—which, by definition, destroys landbases—it is committed to expansion, which means, since someone else already lives there, to conquest. The alternative is starvation. This means the culture must be militarized, with all that implies socially, both internally and externally. I am reminded yet again of Stanley Diamond's famous quote: "Civilization originates in conquest abroad and repression at home." Let's change a couple of words: Civilization originates in agriculture, which requires slavery at home (and abroad) and conquest abroad.

If you base your way of life on the use of a plow, you have to accept the slavery, ecocide, militarization (which also means a high rape culture), and conquest that comes with it.

You could still argue that the fact that humans invented plows shows human superiority over, or greater intelligence than, other species. If agriculture was such a bad idea, you could ask, how has it spread over the earth, until more than 80 percent of the food that humans consume is derived directly or indirectly from plows? This means it's essentially feeding 5.7 billion people. How could I call this a bad idea? Don't I want to eat?

Agriculture has overrun the earth because it provides its practitioners with a potential short term advantage in the application of organized violence. Of course if you convert your landbase into war machines and into soldiers, you will have a short term competitive advantage in a war with a people who don't. This doesn't make you superior, or smarter. It makes you a thief and a murderer, and it makes your way of living unsustainable.

Despite that understanding (or most likely *because* of it) nearly every list of "humanity's greatest inventions" includes the plow. Certainly in the top 100. Almost always in the top fifty. Usually in the top few, along with the wheel, the lever, and the screw. Sometimes it reaches the top of the chart, as being one of the inventions that led to all the rest. As one analysis puts it: "The rise of great cultures and empires was based on plentiful [sic] food supply, and that was based on the plow. Wheat, oats, rye, barley, and other grains could not have been successfully grown without a plow. The plow changed the face of the world and habitat for many of the world's animal species. It was the plow that allowed agriculture to spread across fertile flat lands and push wolves, bears, tigers, and other wild beasts out to the wild and woolliest fringe places of the world."[102]

Please note that they're saying essentially the same thing I am, only they're saying it like it's a good thing. And believing that the invention of plows is a good thing is a big part of the problem.

I'm not sure wolves, bears, and tigers would particularly agree. And I'm not sure those humans who also lived in those "fertile flat lands" until they were pushed "out to the wild and woolliest fringes of the world" would be pleased with being forcibly evicted from their homes. Nor would the fertile flat lands themselves be pleased with being murdered (oh, I'm sorry, reorganized). But to a human supremacist, none of the harm caused by this or any other technology matters. What matters is how the technology helps the supremacist. The point of a supremacist mindset is to facilitate—emotionally, intellectually, theologically, physically—the exploitation of others. If some invention serves that purpose, it is a great invention, and a sign of one's own superiority.

※ ※ ※

Agriculture is usually presented as the solution to food scarcity. My point here is not so much that this is not true, though it isn't. Voluminous literature makes clear that human stature, health, and intelligence all decreased with the rise of agriculture. Diversity of diet decreased. Hunger increased. What agriculture did was allow human population to increase, by converting the entire biome to human use. It also led, as we've discussed, to increased militarization, increased authoritarianism, an increase in rape culture, the destruction of the biosphere, and so on. It is an authoritarian technic, and has led to ever-increasing centralized control of food supplies. Anyone can catch and eat a salmon from the stream, but the walls in the first cities surrounded not the cities themselves, but instead the grain storehouses, not protecting the cities from "raiders" (e.g., the Indigenous peoples whose land the agriculturalists were stealing), but rather the king's grain from the hungry people who might have eaten it and thus not been dependent for their very lives upon their Supreme Leader. Controlling a people's food supply controls them. None of this is the same as being a solution to food scarcity.

But let's pretend for a moment that agriculture *is* simply a solution to food scarcity. Let's compare it to some other solutions, and see which solutions we find more elegant, more helpful, more intelligent, superior.

Some of those living in temperate zones face a food shortage each winter. One approach to this problem is to only live through the summer. This is the approach taken by many annual plants, some insects such as grasshoppers or solitary bees, and many others. Their lives consist primarily of warm and sunny days, as they eat and bask and make love and then leave behind their seeds or eggs for next year. It works for them, and it works for their communities; living only a brief time can sometimes make these plants and animals "first responders" of a sort, who can move in to damaged landscapes and help the land to recover. I'm obviously not suggesting humans (or polar bears) adopt this approach, or even that they *could* adopt this approach. I'm merely saying it's a valid approach.

Another approach is to sleep or doze or drift through the winter. This approach is taken by many deciduous plants, and by many mammals (such as the grizzlies we mentioned earlier), and by many fish. Trees often release hormones into streams telling fish when it is time to rest for the winter, and when it is time to become more active in the spring. Hormones from the trees also act as tranquilizers, and then, come spring, stimulants. Wood frogs freeze solid during winter. Their hearts even stop. In the spring the frogs thaw out, and resume their lives.

A third approach would be to stay awake but eat less through the winter. Many beings do this, from mammals to amphibians to reptiles to birds to plants to fungi and so on. Some humans do it as well. The Algonquin peoples called the full moon in February the "hunger moon," since this would be the month when their food supplies were their lowest. The Cherokee likewise called it the "bony moon." Agriculturalists have often tried to talk Indigenous peoples the world over into adopting agriculture, but most often the Indigenous peoples have understood what would be lost in this adoption, and refused, only to be forced into agriculture through conquest, the elimination of their foodstocks (such as salmon or bison), and other pressures.

Yet another approach is to follow the food. This is what migratory birds do. It is what anadromous fish do. It is what many ungulates do. It is what those who follow the herds of ungulates do. It is what many whales do. It is what many humans do. Even the Tolowa, living here in salmon paradise, still moved up into the mountains in the summer, and down to the coast in the winter. These migrations are wonderful ways to experience different places while you act as nutrient pumps (with anadromous fish, for example, moving almost incomprehensible amounts of food from the oceans into the waters where they spawn). In the case of ungulates and many others, it is a good way to allow land to rest: bison move in, create wallows, and leave for several years as the wallows become homes for aquatic plants, amphibians, reptiles, birds, and so many others. Passenger pigeons brought in and left vast volumes of feces, which decayed into rich soils (the acidity of which also protected the huge chestnut trees where the birds often roosted; the argument has been made that the eradication of passenger pigeons contributed to the devastation of American chestnuts through changing soil composition and making the chestnuts more susceptible to the introduced chestnut blight). The pigeons would stay a while, gift one forest with these nutrients, and then move on to help another. The forests fed the pigeons in the form of nuts, and the pigeons fed the forests in the form of their feces and their bodies. Everybody wins. Or used to, until this human supremacist culture showed up, killed off the pigeons, and nearly wiped out the American chestnuts.

And the final approach we'll discuss here is to store food through the winter. This is part of the promise of agriculture. It does store food, but does so in a way that destroys landbases, leads to hierarchies and militarization, and forces its addicts to continually expand or collapse. Let's contrast that with solutions arrived at by some others to this problem. The Tolowa Indians smoked salmon and jerked meat, and did so sustainably. They did not harm rivers or forests by doing so. In fact, they played similar roles to bears and eagles and ravens and insects and everyone else who eats salmon, in that they carried nutrients in their bodies and then deposited them as feces throughout a forest. This is a vital role in forests, not dissimilar to blood carrying nutrients around the body; it doesn't matter how many nutrients are in your stomach and intestines if these nutrients aren't moved to where they're needed in your body. It's the same in a forest. Or we can talk about honeybees. Honeybees collect food to last through the winter. And their gathering of this food facilitates sexual interactions between flowers. Gosh, we have a solution that leads to ecological destruction and militarism, or one that leads to the exuberance of sexual reproduction and a literal flowering of the next generation. And what is the superior choice?

Or let's talk about squirrels. They're known for gathering and storing nuts in the summer and fall, then throughout the winter, digging up the nuts and eating them. A typical gray squirrel needs about twenty pounds of acorns to make it through a winter. Let's say there are 115 acorns in a pound. That would mean this squirrel would eat about 2,300 acorns in a winter (which, coincidentally or not, is about the same number of acorns produced in a year by a healthy, mature oak of at least some species).

First, since the squirrel hid these nuts, then found them again (partly using smell, but also memory), it clearly has a far better memory than I do. That's a lot of locations to remember. I can't speak for you, but whenever I don't leave my keys in their customary place, I have no memory of where I put them. But squirrels aren't perfect either; they also sometimes forget. They generally find only a little over 25 percent of their caches. Which sounds about fifteen percent better than I would do. In the case of squirrels, this memory loss—or it could be squirrels playing their part in taking care of the forest's future—helps the forest. Squirrels plant far more trees than humans do. And it is simply true that squirrels spend far more time planting trees than they do storing food for themselves; to be clear, squirrels spend far more time taking care of the future of the forest than they do taking care of themselves. I'm sure the trees are more than happy to feed them a quarter of their acorns to thank them for their help. To be accurate at all, the book and film should have been called *The Squirrels Who Planted Trees*, and likewise I probably should have told that simple living dude with the four children that if he thinks not enough trees are being replanted, to take it up with the squirrels. As a side note, squirrels also pay close attention to whether anyone is watching as they hide their food, and if they suspect someone might be eyeing their stash, they'll make decoy caches in which they only pretend to bury acorns. Scientists have also discovered these suspicions extend far past other squirrels; when the squirrels realized the scientists were disturbing their food supplies, they started making more fake caches to throw the scientists off the trail, or at least waste their time.[103]

All of which is a long way of asking, which is a better way of storing food for the winter: one in which you deplete the topsoil, destroy the landbase, create and support authoritarian power structures, then conquer other landbases and destroy them, too; or one in which by your very act of storing food for the winter you guarantee the health of your home and its future for your own children and those of the other species who share this larger body that is the biome?

The squirrels are helping out the trees, who are helping out the squirrels, who are helping out the trees. . . . What did Richard Dawkins call beings who acted like this? Oh, yes, Suckers.

What do I call them? Life being life.

It would be easy enough to do what I've done so often, and simply make a snide comment about who is more intelligent between human supremacists and squirrels (I'd say the squirrels, since they generally hide the nuts, instead of valorizing them as respected philosophers), but the first point I really want to make here is that both nonhuman and human cultures have come up with a near infinitude of "solutions" to this particular "problem." So it's nonsense to say that humans—or let's just say what we mean and say non-Indigenous humans—are superior because of our ability to solve problems. Further, these other solutions have had the necessary elegance of not only not destroying their landbases—what they rely on to, you know, live—but rather improving their landbases, all of which would seem to me to be the number one consideration for whether a solution is or is not superior.

But there's a deeper problem here. When human supremacists talk about human superiority being based on human capacity for problem-solving, or innovation, or technology, or even epistemology or religion, nearly always the exemplars of this superiority are not merely clever pieces of technology, but are instead authoritarian technics.

Recently *The Atlantic* "assembled a panel of 12 scientists, entrepreneurs, engineers, historians of technology, and others to assess the innovations that have done the most to shape the nature of modern life. The main rule for this exercise was that the innovations should have come after widespread use of the wheel began, perhaps 6,000 years ago. That ruled out fire, which our forebears began to employ several hundred thousand years earlier. We asked each panelist to make 25 selections and to rank them, despite the impossibility of fairly comparing, say, the atomic bomb and the plow."[104] Given the destructiveness of each, I'd say comparing the atomic bomb and the plow is dead easy. Both are weapons of war. And here's a simple comparison: the amount of energy used for agriculture each year in Iowa (including the energy used to create fertilizers and pesticides) is equivalent to 4,000 Nagasaki bombs.

They called their list, "The 50 Greatest Breakthroughs Since the Wheel." Let's start with a caveat: many of their entries weren't actually human breatkthroughs or innovations at all. For example, they listed penicillin, anesthesia, nitrogen fixing, and electricity as human innovations, which I'm guessing would cause fungi, plants, other plants, and matter itself to do the equivalent of shrugging and muttering, "Fucking typical." It shouldn't surprise us that they didn't credit nonhumans. They didn't even credit Indigenous peoples. I'm sure the Indigenous humans who have had long relationships with coca, poppy, and cannabis plants, among many others, would be surprised to learn that anesthetics were invented in 1846.

Now, to the point: essentially all of the "greatest breakthroughs" were authoritarian technics, and have had the effect of increasing the ability of those in power to exploit. For example—and remember, they are listing these as the "greatest breakthroughs," not "the most horrible inventions"—number thirty-two was the cotton gin, because it, and this was the entirety of the reason given for its inclusion, "Institutionalized the cotton industry—and slavery—in the American South." I'm not sure enslaved persons would have agreed that this was one of the "greatest breakthroughs" in history, any more than German Jews circa 1942 would have agreed that the technics involved in death camps were a great breakthrough. And please note that some of the "scientists, entrepreneurs, engineers, historians of technology, and others" involved in making this list said explicitly that they chose what innovations to include by asking themselves, "What would I miss more if it didn't exist?" Gosh, would I have missed cotton gins? If my loyalties were with the slaveowning class, then certainly yes. If my loyalties were with the enslaved, then of course not. Please note also that although the write-up made some obligatory noises about how "progress" might carry with it some potential downsides, the overall tone of the write-up was self-congratulatory and "optimistic." Indeed, the most significant part of the exercise, according to its author, was to help us understand much about "why technology breeds optimism." Do you think the invention of the cotton gin engendered optimism among the enslaved? The author states in the introduction that the list is intended to reveal much to us about "imagination, optimism, and the nature of progress." I think he's right; the list does reveal almost everything we need to know about imagination in a culture in thrall to authoritarian technics, and what optimism means in a culture based on slavery, and about the nature of progress in this culture; just not in ways the listmakers meant. Another of the "greatest breakthroughs" was oil drilling, about which these "optimists" stated that it "fueled the modern economy, established its geopolitics, and changed the climate." Once again, I'm not sure victims of the modern economy, those whose blood has been spilled for oil, or victims of global warming would be eager to jump on this bandwagon. Yet another breakthrough was radio, because it was "the first demonstration of electronic mass media's power to spread ideas and homogenize culture." In other words, one of their "greatest breakthroughs" is a tool for those who own the media to spread propaganda and reduce or eliminate cultural variation. Of course another of their "greatest breakthroughs" is television, which "brought the world [sic] into people's homes," by which they really meant: "The most effective propaganda tool yet devised for spreading the ideas of those in power." But from within a supremacist worldview, "the world" actually already means "the ideas of those in power." No more, no less. That is what the world must consist of. Another was the assembly line, because it "turned a craft-based economy into a mass-market one." Yes, that same assembly line that was part of the efficiency movement about which Frederick Winslow Taylor famously wrote, "In the past the man has been first; in the future the system must be first." You couldn't really ask for a better description of what authoritarian technics do to a society, and of what has gone wrong with this culture. Number twenty-one was nuclear fission, because it "Gave humans new power for destruction, and creation [sic]." I'm sure I'm not the only person who, when human supremacists extol nuclear power, finds himself thinking of *The Sorcerer's Apprentice*, and Mickey Mouse's misuse of the spell to harness the power of broomsticks. I find myself thinking that, like Mickey Mouse, humans are creating a terrible mess by using this power they can never understand or control. Unlike Mickey, however, because of this error we (and everyone else) will receive much more than a swat on the butt. And we will deserve it. The other members of the planet will not.

And the fourteenth "greatest breakthrough" was gunpowder, because it "outsourced killing to a machine."

Think about that.

I hope you're beginning to see the pattern: each of these is an authoritarian technic. Each of these has among its effects the centralization of power. Each of these is useful from the perspective of those who are interested in increasing power at the expense of nonhuman and human communities and communal variability. Most of those that are not strictly authoritarian technics are supported by counterfactual claims, such as number 47, the nail, which they say "extended lives by enabling people to have shelter." Really? Nobody had shelter before nails? Indigenous peoples the world over have never had shelter? Huh? But the real point is that in none of the cases, not even anesthesia, which I would call a good (except, of course, that, just as with electricity, penicillin, nuclear fission, and nitrogen fixing, humans didn't invent it), do the listmakers count the cost of these great "breakthroughs." For crying out loud, they don't seem to be particularly broken up about the costs of gunpowder, nuclear fission, and the cotton gin, so why would we expect them to care about the fact that the production, consumption and excretion of birth control pills (number twenty-two) has caused terrible hormonal changes among fish who live in rivers where the hormones eventually end up? Why would we expect them to care about the costs of any of these inventions? Why would we expect them to care that the overwhelming majority of these innovations have been terrible disasters for the real world?

In fact, in most of the lists like this one, a primary precondition for some innovation being considered one of the "greatest breakthroughs" is that the "breakthrough" consolidate power, that it further enslave nonhumans (and humans), that it make ever more matter and energy jump through ever more hoops on command. That it centralize control.

When you begin to question human supremacism, such that it no longer is the real authority of your own mind, these lists begin to look much different than they did before. They seem to be lists made by those who don't consider themselves to be part of the earth. Instead, these are lists compiled by those who are at war with the world itself, with life itself. Suddenly it makes sense that plows and gunpowder and atomic bombs are all included in a list of greatest innovations.

CHAPTER ELEVEN

BEAUTY

Art is implicit in nature.
ALBRECHT DÜRER

Everything has beauty, but not everyone sees it.
CONFUCIOUS

Here is my own informal list of what I think are a few really great innovations, inventions, and creations, this from a perspective that is at least an attempt to not be human supremacist, and that does not take as a given that what humans create has meaning and what anyone else creates does not. I'm not going to claim that these are the most foundational creations, or most important, or anything else other than that they are pretty great.

There is matter, space, and time. Without them there is nothing.

There is nuclear fusion, developed by the sun and other stars. Without it we'd all be very cold.

There is gravity, and other forms of attraction. Without it things would fall apart rather quickly.

There is motion, developed by the first entities who moved.

There is electricity, which was not in fact developed by humans.

There is sunshine, which feels really good on a nice fall day. Don't you love how it warms you all over?

There is homeostasis. How great is that?

There is fire, which also was not developed by humans, but by fire itself.

There is water, and there is ice, and there are clouds. There is rain.

There are oceans. There are rivers. There are springs. I remember as a child marveling at a huge bubbling spring that birthed a river fully formed, and wondering how it never ran out of water. There is the whole hydrologic cycle.

There are rocks, like water their own beings, in many cases long lived. They are foundational.

There is metabolism. Eating is a good thing, is it not?

There is cell division.

There is oxygen combustion.

There is sexual reproduction. There is reproduction without sex. There is sex without reproduction.

There are butterflies. Moths. There are leaf insects. There are grasshoppers who bury their clutches of eggs, like tiny sea turtles.

There are fireflies.

There are the fall colors of trees.

There is that light green of new growth on trees.

There is the sound of feet on dry leaves on a forest floor.

There are feet.

There are cilia.

And there are eyes. Can you imagine anything so brilliant? Who came up with that?

Or what about the sense of touch? Is this more brilliant than vision? Smell. Hearing. Taste. Which is the most brilliant? Don't ask me; I'm certainly not smart enough to figure it out. They're all good. And what about other senses unknown to humans?[105]

Fruits and berries. One of the most brilliant ideas ever: putting your seeds in attractive, nourishing packaging which will lead someone to consume your seeds, deliver them elsewhere, and plant them in a bed of manure, which, it ends up, is, in another burst of brilliance, food for your child. Everybody wins! Is this a great idea, or what?

Proprioception. Have you ever wondered how hard life would be if this innovation had never taken place?

And while we're talking about bodies, isn't medicine amazing? Of course, humans didn't invent the practice of the "diagnosis, treatment, and prevention of disease."

There are anglerfish and whale sharks. There are algae.

Spider silk, echolocation, beaver dams, birds' nests, flowers.

Color.

Lichen.

Pheromones.

There are families of wolves, families of baboons, families of elephants, families of chimpanzees, families of bees, families of alligators, families of frogs, families of plants, families of bacteria, multispecies families, like forests, like rivers, like you. There are families.

Friendship is a wonderful innovation.

Love.

There are roots to nestle deeply into soil. There are roots who wrap around each other to hold up friends and comrades and lovers.

Wings. There are wings for flying and wings for swimming.

Fat. Isn't that a marvelous way to store energy and to keep you warm?

Muscle. Who invented muscles? They're extraordinary.

Blood. Sap. Water.

Dreams.

Silence.

Beauty.

Harmony.

Anger.

Sorrow.

Joy.

I could go on and on, but perhaps it's best if you come up with your own.

❂ ❂ ❂

It also bothers me that human supremacists believe that only humans create art.

What about lightning?

Thunder?

Ocean waves. Their sound, their smell, their sight.

Have you ever seen seals body surfing?

Clouds.

Toroidal bubbles.

Non-toroidal bubbles.

The wind in the trees, from soft sighs to groans to creaks to the clashing of branches. Don't you think a light breeze in a deciduous forest is the best sound in the world? Or no, maybe that would be a heavier wind in redwoods. Or maybe a full-blown storm in a pine and fir and cedar forest, where storm and forest together sound like the sea.

Or maybe it's the sound of meadowlarks.

Or maybe an interspecies chorus of frogs.

Or maybe the sound of a herd of bison running as fast as they can.

Or maybe the hundreds of songs of mockingbirds.

Or maybe the sound of a cave breathing.

I love the four seasons. No, not the cover version by Vivaldi, much as I like it, but the original. The original came *much* earlier, and is *much* better.

Snow.

Fog.

Frost.

Larches in fall.

Maples in fall.

Willows in spring.

The look in a dog's eyes when it's happy.

A deer's eyes. The eyes of a jumping spider.

Octopi, some of the world's best actors.

Snakes. Aren't they beautiful?

Egrets.

Alders.

Canyons.

The wings of a dragonfly.

Iridescent beetles.

Non-iridescent beetles.

Skin.

Jellyfish tentacles.

Chanterelles, Amanita muscaria, earthstars, bridal veil stinkhorns, puffballs.

Murmurations of birds or fish.

Snowflakes.

Raindrops.

Water-sculpted rocks. Wind-sculpted rocks.

The smell of a forest. Is there anything more beautiful? I don't think so.

The smell of a desert after a rain. Is there anything more beautiful? I don't think so.

The Milky Way. Is there anything more beautiful? I don't think so.

Agates. Crystals. Granite. The smell and feel and sight of soft soil.

Sunrises and sunsets.

How about the first faint star you see after the sun goes down, hanging just so above the silhouette of a ridge of conifers black on deep, deep blue? Is there anything more beautiful? I don't think so.

❀ ❀ ❀

Life itself is art.

By destroying life, human supremacists are destroying the most beautiful and extraordinary innovations, creations, art.

How could they do this?

They value only themselves. What humans create has meaning. What nonhumans create does not.

Not only must human supremacists devalue what nonhumans create, they must destroy these creations, lest these creations remind them that there still exist those who are not under their control, lest it remind them they are not the only beings who exist, that they are not the only beings who create.

This is one reason this culture is so destructive. It is one reason it hates nature so much. The real world keeps reminding us that life is not all about us.

CHAPTER TWELVE

CONQUEST

Sometimes people talk about conflict between humans and machines, and you can see that in a lot of science fiction. But the machines we're creating are not some invasion from Mars. We create these tools to expand our own reach.

RAY KURZWEIL

We see hatred of nature everywhere in this culture. And I mean everywhere. Tonight I saw an op/ed in *Forbes Magazine* entitled, "In the Battle of Man vs. Nature, Give Me Man." The article begins, "Welcoming the new year contemplating the sunset comfortably ensconced on a cliffside balcony high above the manicured banks of the Miami River, it's hard not to marvel at the hand of man. Behold as lights defeat the growing darkness, lending sparkle to a condo canyon that was once a malarial swamp. Yes, the pristine wilderness is a wonderful place to visit, but most rational people would rebel if forced to live there."[106]

There are, of course, many things wrong with this, not the least of which is that the "battle," or rather war, or rather massacre, being waged by "Man" against the rest of the world—a.k.a. "Nature"—is killing the planet. Next, of course, is the insanity of the belief that you can win a war against the planet that provides the basis for your own life; or more accurately, the insanity of the belief that winning a war against the planet that provides the basis for your own life can end in anything other than your own demise as well as the planet's; where does he think the raw materials come from to build these condo canyons, and where does he believe the energy comes from to power those lights? More importantly, where does he think food and water and oxygen come from? Of course what "winning" this war would look like to him and people like him is not the murder of the planet—you can't perceive yourself as murdering something you perceive as already inanimate—but rather its complete bending to his will. Its "reorganization." Next, his preference for the artificial over the natural, in this case city lights to night (and moonlight, starlight, or darkness) and condominiums to wetlands; and his near-Biblical and certainly narcissistic reverence for "the hand of man," are not only measures of this culture's sickness, but more basically are pretty straightforward statements of common beliefs that *are* this culture: that the enslavement of the world is a good thing, and that this enslavement is possible without murdering the planet.

I was also bothered by this statement: "Yes, the pristine wilderness is a wonderful place to visit, but most rational people would rebel if forced to live there." First, until only a few thousand years ago (and on the Miami River, until only a few hundred years ago), what he calls "pristine wilderness" was not called "pristine wilderness," and it wasn't a place for people to visit. It was called "home," and it was where people lived, people who fought against the conquest and enslavement of their homes, people who did prefer wetlands and starlight to condominiums and city lights. Also, saying that "most rational people would rebel if forced to live there," implies that those who gladly lived there were not as rational as those who destroyed these "wildernesses" and the humans (and nonhumans) who called these places home. It implies they were not as rational as those who live in condo canyons. This is fully in line with the disturbingly common belief among members of the dominant culture that Indigenous peoples—a.k.a. people who live in "pristine wilderness," a.k.a. "primitives"—are too often not perceived as fully rational.

I'll tell you what is not rational, or reasonable: harming the capacity of the earth, our only home, to support life. Nothing could be more unreasonable or irrational or stupid or evil than that.

I want to mention one more passage, from near the end of this *Forbes* essay: "Will we give a clear mandate to leaders who celebrate man's exceptionalism, understanding that the incidental problems created as we harness technology to bend nature to our will can be solved using more technology? Or will we cede power over every aspect of our lives to an elite [sic] that claims to speak for the inanimate [sic] environment and seeks to command us to live with less, redistribute our property, and empower

politically appointed central planners to scale down and reshape civilization to appease Mother Nature's wrath?"

Here we go again, with human exceptionalists, which is really just
another name for human supremacists—and the same is true for white
exceptionalism (or supremacism), male exceptionalism (or supremacism), US or capitalist or civilized exceptionalism (or supremacism)—
dismissing the harmful effects of their exceptionalism and supremacism.
As always, this dismissal happens because the harmful effects are suffered
by the victims of the supremacists, and not generally the supremacists
themselves, who are then—what a surprise—generally the beneficiaries
of the exploitation that follows from this exceptionalism or supremacism. Two hundred species went extinct today. Ninety-eight percent of
old growth forests are gone. Ninety-nine percent of prairies. Ninety-nine
percent of wetlands. Ninety percent of the large fish in the oceans are
gone. Shellfish in the Pacific Northwest are undergoing reproductive
failure because of industrially-induced acidification of the oceans. And
these are what he calls "incidental problems," that is, when he doesn't
claim they're positive goods. And remember, he is not the point; the
point is that he's articulating a destructive and narcissistic attitude that
is the dominant culture—that the extirpation of nonhumans is at most
an incidental problem, but more likely either progress (converting nasty
swamps to glorious condo canyons), production (developing natural
resources), or something completely inconsequential. Because it's happening to someone who—or, in the human supremacist formulation,
something that—is not fully alive, not fully "rational," not fully aware,
and certainly not worthy of moral consideration.

Sometimes the extirpation of nonhumans is perceived as "saving the
earth," as in an article in today's *Los Angeles Times* headlined, "Sacrificing
the Desert to Save the Earth."[107] The article is about how state and federal
governments, a big corporation, and big "environmental" organizations/
corporations are murdering great swaths of the Mojave Desert to put in
immense solar panels. The desert is being sacrificed not, as the article
states, to save the earth, but to generate electricity, primarily for industry.
The earth doesn't need this electricity: industry does. But then again,
from this narcissistic perspective, industry *is* the earth. There is and can
be nothing except for the supremacists themselves.

Here are a few of the other problems with this *Forbes* text, problems which are nearly universal in this culture's way of being in the world. First, there is the immorality (and, in this culture, the ubiquity) of wanting to "bend nature to our will." Or we could talk about this writer for *Forbes* waxing enthusiastic about bending the entire planet to (his perspective of) "our will," and then immediately afterwards accusing someone else of being part of some elite. Uh, wouldn't wanting to bend the world to your will make you by self-definition part of an elite? Or we could talk about the cognitive dissonance that inevitably follows when we propagate lies like human exceptionalism, in this case the dissonance manifesting as calling nonhuman nature inanimate, but then immediately speaking of "Nature's Wrath." Which is it: is "Nature" inanimate or is it wrathful? You can't have it both ways. And of course his language reveals that on some level even this human supremacist recognizes that the real world has reason to be wrathful.

I find it extraordinary—and of course, entirely expected—that so many human supremacists speak blithely of bending the entire world to "our" will, and attempt to force all of us to live with less of the planet (and to force all those exterminated to not live at all), but then they freak out at the possibility of anyone in any way constraining any of *their* own freedoms, at the possibility of someone "commanding" *them* to live with fewer luxuries (luxuries that are gained by forcing others to bend to their will), and freak out as well at the possibility of reshaping this culture to be in line with the needs of the planet, the source of all life.

* * *

Here's another fairly typical argument: the plow was the greatest invention of all time "not because it makes all else [sic] possible, but because it single-handedly diverted the direction of the human race to a wider degree than anything else." Or, "I have heard it argued convincingly that the greatest invention ever was the plow. It allowed us to have surplus food, which allowed armies, priests, scientists, builders, just about everything [sic]." Just about everything, that is, except peace, which it makes impossible; and justice or even survival for those about to be conquered or exterminated; and sustainability, which, like peace and justice and the survival of the victims, was for this culture never even a consideration.

Please note again that it's just plain wrong to say that the plow "allowed us to have surplus food." Don't you think an entire river full of salmon is more food than local humans (or bears, eagles, ravens, trees) could eat? Why doesn't that qualify as "surplus food"? Prior to the plow, the world was already full of food. It just wasn't under human control, or more precisely the control of an elite. It was available to humans and non-humans, without regard to any individual or collective human wealth. This means that within this culture that is based on authoritarian technics, not only won't these wild food surpluses be considered real—the only real food surplus, like the only real meaning, is one humans create and control—but worse, that these other communities that provide these food surpluses must be eradicated in order to maintain control of human populations; how are you going to force people to work for you if they can find food, clothing, and shelter on their own? All of this means that, as is true for innovations, food (or other) surpluses that contribute to democratic social structures will be undervalued, privatized, exploited, and destroyed. Food (or other) surpluses that contribute to authoritarian social structures will be lauded as innovations, cultivated, and controlled.

So, if you think the diversion of much of the human species into a direction that is ultimately going to kill the planet (but allow the richest of humans to have lots of "comforts or elegancies" in the meantime (while their human and nonhuman slaves lead lives of grinding immiseration)) is a good thing, the plow is your invention. Likewise, if you think armies, priests, and scientists are good things on their own, or in any case are worth more than the liberty and lives of all those harmed by the entire agricultural technics, then the plow is for you.

❁ ❁ ❁

You could argue that it doesn't matter how destructive and disastrous plows and agriculture (or other authoritarian technics) have been for the entire planet. They have helped human populations to expand, and they have helped "push wolves, bears, tigers, and other wild beasts out to the wild and woolliest fringe places of the world." That by itself means we're smarter and superior; were we not smarter and superior, we would not have been able to conquer and exterminate them. They would have conquered and exterminated us. In this sense, far from arguing that the destruction of wild places doesn't matter, the argument would be that this destruction—this conquest, this transformation—is actually a sign, if not *the* sign, of our intelligence and superiority. Which is the real point, and has been all along. It's also, ultimately, the argument that underlies and is the real reason for all of the other arguments for human supremacism.

Both intelligence and superiority are here conflated with conquest and murder. But that only works if your definition of intelligence or superiority means not only acting atrociously—might makes right; might makes intelligence; might makes superiority—but also greatly decreasing the capacity of the planet to support life. By which I mean not only non-human life—which, at best, doesn't count to human supremacists, and often is considered pestilential—but human life, as well. I know there are a lot of humans alive now, but what do you think will be the human population when the oceans are dead?

Recently Richard Dawkins said he believes humans have a 50 percent chance of surviving this century. The tools of science, he says, have enabled scientists to create weapons powerful enough to kill all humans; his fear is that religious fundamentalists will get ahold of these weapons and use them (never mind what capitalists already do with the weapons science has provided for this culture's war on nature). If we choose as our "*sine qua non* of behavioral intelligence systems" "the capacity to predict the future; to model likely behavioral outcomes in the service of inclusive fitness," would creating tools that are powerful enough to destroy life on the planet—or at the very least, all humans—not, in all truth, disqualify us from being considered intelligent? Actions leading to a realistic chance of driving your own species extinct (and taking down much, if not all, of the planet in the process) clearly are not "in the service of inclusive fitness."

Dawkins is not alone in perceiving humans as causing their own near-term extinction. Stephen Hawking has famously remarked that in order to keep from driving ourselves extinct, humans need to colonize space.[108]

The real point, apart from Hawking's appalling and sociopathological—and completely typical for this culture—lack of concern for everyone else on the planet, is that even though he understands that human behavior is killing the planet, he refuses to question human supremacism, or the *right* of humans to murder every known living being in the universe. Him and just about everyone else in this culture.

A couple of years ago a mechanistic scientist said to me, "The miraculous explosion of knowledge these past few centuries since the industrial revolution is almost—*almost*—worth the cost in terms of environmental destruction."

I was horrified to hear this, not only because he ignored the knowledge lost as this culture eradicates Indigenous human and nonhuman cultures—as scientific knowledge and power have increased there has been a consequent and easily predictable decrease in other forms of knowledge, such as, for example, that knowledge held by and contained in passenger pigeons and the humans and nonhumans who relied on them—but also because of his clear expression of a human supremacist perspective; I'm guessing that passenger pigeons and the forests who depended on them would not so readily agree that their own eradication has *almost* been redeemed by the increase in scientific knowledge and power wielded by industrial humans.

❊ ❊ ❊

Pretend I run a business. Let's say I make doughnuts. My store is called Doughnut Supreme—Latin name *doughnutus supremus supremus*—and that name is entirely deserved. How do we know it's deserved? Because I say it is. I write lots of reviews extolling the supreme quality of my own doughnuts. I develop a religion called The Church of the Supreme Doughnut where the Giant Baker in the Sky (who looks remarkably like the baker in my logo) describes how my doughnut shops are supposed to go forth and multiply. The first commandment of this religion is, "Thou shalt have no other bakers before me." I develop an epistemology that declares we know something to be true if it begins with the understanding that my doughnuts are the best. I develop a literature in which the heroes run Doughnut Supremes, and the purveyors of other doughnut stores are either nonexistent or are obstacles to be overcome: "In the battle between Doughnut Supreme and everyone else, give me Doughnut Supreme." I love to make long lists of Doughnut Supreme's greatest innovations. I propagate the notion that if Doughnut Supreme makes a doughnut, it is a *doughnut*, full of meaning and import. Any other "doughnut" made by anyone else is not a true doughnut, and does not serve the functions of true doughnuts.

I know my doughnuts are superior, and my store is superior, not only because I say so, but also because Doughnut Supreme is the most profitable doughnut store in the world. And because profits are a central measure by which every endeavor must be judged, Doughnut Supreme is supreme! It expands across the world. There are Doughnut Supremes everywhere! I am, by *any* measure, the most successful and intelligent doughnut chain owner on the planet. Which means I am the most successful and intelligent being on the planet.

So, how did I get to be so successful? How have I been able to expand across the globe? How have I been able to drive every other "doughnut" store out of business? It's because I'm the best! That's how.

That's where you come in. You say to me, "It's very simple, really. You never pay your bills. You don't pay rent. You don't pay for materials. You don't pay for energy. You don't pay for labor. Of course if you don't pay your bills you're going to run a profit. You'd have to be an idiot not to, right?"

"No, not true. I run Doughnut Supreme, and Doughnut Supreme has overrun the planet, which means I can't be an idiot. I must be supremely intelligent. Otherwise Doughnut Supreme wouldn't control the world! In fact, I have just declared that the modern geological age should be called the Doughnutsupremocene, because Doughnut Supreme has become a world-shaking geological force!"

"But," you say, "You don't pay your bills."

To which I reply, "Now wait a BakerDamn minute. We at Doughnut Supreme are responsible stewards and responsible businesspeople. We pay our bills. Of course we don't pay rent, because we *own*. Actually a good part of the world by now. But we pay rent to ourselves. And we pay for the finest wheat, straight from what used to be the prairies, and the finest sugar, from what used to be the Everglades. And we pay for electricity and labor, too! How dare you accuse us of not paying our bills!"

"Whom do you pay for electricity?"

"It's green energy from a hydroelectric company, which I also happen to own, through different corporations. But let's leave that aside . . ."

"Whom did the hydroelectric company pay for the electricity?"

"They paid to build the dams (with cement, number thirty-seven on *The Atlantic's* list of 'greatest breakthroughs,' which, it says, is 'at the foundation of civilization as we know it—most of which would collapse without it'), and they paid to build the electrical grid (well, actually, the government paid for that, but leave that aside, too, and now that I think about it, the government also paid for the dams). They pay their workers, and so on. It's all paid for, one way or another. And besides, why do I care who the electric company pays?"

"Where did the energy come from?"

"I just told you, the dams."

"No, the dams convert the energy into electricity. Where does the *energy* come from?"

"The river, I guess."

"And who paid the river?"

"Don't be stupid. A river can't use money."

"So give the river what it does want."

"Which is what?"

"Ask the river."

"Don't be stupid."

"Let's try this again."

"What?"

"Whom did you pay for wheat and sugar?"

"Farmers. Well, actually huge agricorporations. But same diff, right? Oh, and I own them, too, but . . ."

"But yes, we'll leave that aside. . . . So, whom did the agricorporations pay?"

"Chemical companies, and the bank—"

"—Which you also . . ."

"Yes, and . . . ?"

"Who grew the wheat and sugar?"

"I just told you, agricorporations, I mean small independent family farmers."

"No, who grew it?"

"Aren't you listening?"

"Who paid the soil? Who paid the wheat and sugar cane plants? Who paid the prairies? Who paid the Everglades? Aren't they the ones who actually grew it?"

I say, "What the hell are you talking about, you crazy person?"

You say, "Doughnut Supreme is overspreading the planet because you don't pay your bills. You are, to use Richard Dawkins's term, a Cheat."

What follows is an awkward silence while everyone in the room politely forgets you said anything at all.

* * *

You could also argue that it's not the human invention of plows (and other weapons of mass destruction) as such that implies human supremacy—after all, murdering the planet isn't really that great of an idea—but instead it's the human *capacity* to invent plows that implies this supremacy. I mean, neither squirrels nor blue whales nor redwood trees nor shrews (with their 10 percent brain-mass-to-body-weight ratio) nor fungi (with their essentially one-to-one brain-mass-to-body-weight ratio) have been able to build plows, or bicycles for that matter. No species other than humans have been able to invent plows, Pop-Tarts, two-stroke engines, intercontinental ballistic missiles, computers, or skyscrapers. And the truth is that we really are very good at creating gadgets. It is perhaps our most obvious gift as a species, this creation of gadgets. But there's another problem. Does the capacity to invent gadgets really imply superiority, or even intelligence? By now we can see that the implication would be at best tautological. But here's a far worse problem: if this gadget-making somehow becomes so compulsive that the gadgets threaten your own survival and the survival of the planet, such that even such a science booster as Richard Dawkins can acknowledge a 50 percent chance of these gadgets eradicating all humans within the next eighty-five years—and *still* not question the compulsive creation of ever more, and ever-more-dangerous, gadgets—how can even the most serious proponents of gadget-making still presume that the capacity to create gadgets is a sign of intelligence or superiority? The best we can say is that it sounds like a serious problem to be resolved. But honestly it sounds more like a terrible addiction.

We also cannot forget the cultural component to this gadget-making. The Tolowa certainly invented some gadgets, such as baskets and hand-woven fishing nets, but they never allowed their gadget-making to become so compulsive as to cause them to destroy their landbase. As a consequence, they were able to live here for at least 12,500 years without destroying the place. The dominant culture has been here for 150 years, and the place is trashed.

Why would we do something so stupid as to invent gadgets that threaten our existence and the existence of life on this planet? And why would we presume this means we're superior? The answer: because our self-perceived superiority is *based* on our ability to enslave or destroy others.

And of course, all these questions are necessarily linked. Addicts often fail to recognize their own addictions, and perceive themselves instead as making choices. And it's just some sort of strange coincidence that the addicts' choices happen to consistently feed their addictions. They can quit any time they like. Maybe tomorrow. Or the day after.

But there's a huge and fatal difference between an alcoholic opening another bottle, and members of the dominant culture creating more effective means to "make matter and energy jump through hoops on command, and to predict what will happen and when." The difference has to do with the fact that those who are addicted to power and control receive tangible benefits for feeding their addictions. This is one reason perpetrators of domestic violence—and slavers, and capitalists, and exploiters in general—rarely change: their behavior is gaining them tangible benefits. Never mind that doing so harms their relationships; if they're addicted to power and control, making others jump through hoops on command is by definition more important to them than having loving mutual relationships. The cliché is that addicts don't usually change till they hit bottom. But those who are addicted to power and control are not the ones who hit bottom; it's their victims who hit bottom. These particular addicts will not change so long as there is any other option, and quite often, not even then.

CHAPTER THIRTEEN

THE DICTATORSHIP OF THE MACHINE

Men have become the tools of their tools.
HENRY DAVID THOREAU

All of which brings us back to Lewis Mumford. The fact that an authoritarian technics emerges from and leads to authoritarian social structures is only part of why that technics is called *authoritarian*. Another, perhaps more important reason has to do with how the technics themselves gain authority over a culture. The logic behind the technologies can come to rule. The technics, and not the people, and not the landbase, are in control.

We see this all the time. Or more precisely, because unquestioned assumptions are the real authorities of any culture, we don't *see* this; it rules our lives, but we take it as normal.

For example, let's talk about fracking. Fracking is sold as a way to get more energy. More energy is sold as way to make people's lives better. Among many other problems, fracking poisons groundwater. Communities are having to fight to protect the water they drink. As in drink. As in one of the things we have to do to survive. This means that those who benefit financially from fracking are poisoning the groundwater necessary for the survival of affected community members. A reasonable descriptor for those doing this is *sociopath*. Not only must sociopaths be stopped, we also need to ask what is wrong with a society that allows sociopaths to poison the drinking water of members of other communities. No, it doesn't *allow* sociopaths to poison groundwater. It encourages them to do so, and rewards them for this behavior.

The frackers can (and will) argue that they are doing what is best for the economy. And that will be making my point precisely: who is in charge? Who is actually making the decisions? Are they being made by human beings in community, or are they being made by those who are serving the technics that called fracking into being, and which is being influenced by that same technology? The technics is controlling society, causing it to poison even its own groundwater.

Two days ago a judge overturned a ban on fracking voted in by the people of Fort Collins, Colorado, writing, "The City's five-year ban effectively eliminates the possibility of oil and gas development within the City. This is so because hydraulic fracturing is used in 'virtually all oil and gas wells' in Colorado. To eliminate a technology that is used in virtually all oil and gas wells would substantially impede the state's interest in oil and gas production."[109]

There you have it. Neither protecting your drinking water from being poisoned nor any notion of community self-determination shall be allowed to impede oil and gas production.

Remind me again: who's in charge?

We can do the same exercise for oil. Same selling points. Same harm. And we can add its role in murdering—sorry, reorganizing—the planet. The planet is undergoing the most rapid heating in its history, contributing to the greatest mass extinction in its history, and a fair number of people believe global warming will drive humans extinct within the next generation or two. Yet this society keeps on exploring for, extracting, refining, and burning oil. Is it just me, or does this line of action seem to have a very strong downside? Once again, who is making these decisions? Either sociopaths who must be stopped, or the technics itself, which must be dismantled and destroyed.

We can go through a whole raft of other technics, but it all boils down to the same: actions are taken to protect and further the technics, not living beings. We can do this for corporations. Corporations are ostensibly legal tools to facilitate commerce. But when corporations—legal fictions—control social decision-making processes, the tools are literally in charge. The tools—corporations—are authoritarian. We can do this for money. Money is ostensibly a legal tool to facilitate exchange. But when social decisions are made not primarily because they serve humans and nonhumans, that is, not because they serve life, but rather because they "make money," then money is obviously controlling or guiding these decision-making processes. This is true on smaller scales, as individuals are forced to make decisions they would not otherwise make, because they're forced to earn money to survive in a capitalist economy (which, as we've discussed, is not coincidental; the laws of apartheid, for example, were drafted specifically to drive people out of their subsistence economies and into mines). And this is true on larger scales, as the wealthy often have far more money than they will ever need to survive the rest of their lives, and still they continue to accumulate; money has become an end in itself. We can do this for power. When social decisions are made not primarily because they serve life, but because they increase the power of the decision-makers and others of their class, then power itself and not life is the real authority behind the decisions. We can do this for agriculture: once you have set yourself on the path of overshoot and drawdown—*overshoot* is when a population of any given species living in a particular manner exceeds the place's carrying capacity (or the maximum population of that species who could live in that place in that way forever without harming that place); and *drawdown* is the harm done by these overpopulations who exceed carrying capacity, permanently drawing down carrying capacity—the technics itself and the physical conditions it creates lead to conquest and slavery. This can only stop with the (probably involuntary) abandonment of the technics. We can do this for "technological progress," which is more accurately termed "technological escalation," since the real point of the "progress," as we've seen, is most often to escalate the control and reach of those in power. This is entirely to be expected in a culture based on authoritarian technics. It is also to be expected in a supremacist culture. And if you have a culture based on competition—and of course, it often comes as a complete surprise to members of this highly supremacist, highly competitive culture to learn that there have been cultures who are neither supremacist nor competitive; and to learn, further, that the erroneous belief that *every* culture, indeed all life, is and must be guided by competition is itself a central social part of an authoritarian technics—that competition will drive this "progress," this "advancement," this escalation.

If, as in Dawkins's story of Suckers and Cheats, you have two cultures who are not supremacist, not based on authoritarian technics—in other words, they are, to use Dawkins's word, *Suckers*—they can coexist more or less forever. Now introduce a third culture, which believes in the Great Chain of Being, which perceives itself as superior to these others, which is based on authoritarian technics, which, through overshoot, has converted its landbase into human beings (and, most importantly to this particular example, soldiers) and into machines for war. What happens next? Well, that's a story we've seen a few times over the past several thousand years. The authoritarian culture will do its worst to wipe out the non-authoritarian culture and steal their land. It will then proceed to steal from and destroy—I mean, manage; I mean, reorganize—this landbase to fuel its authoritarian structures and to fuel further conquest. Those survivors among the non-authoritarian cultures who aren't wiped out will probably, if they are to continue to survive among the Cheats, need to adopt at least some of the attributes of the authoritarian, conquering culture. Now let's introduce a fourth culture, which is also supremacist, authoritarian, and so on. Let's say the machines of war of the two empires are on par. Next, one of them invents a new technology of killing (or otherwise extending the control of those in power). What happens then? The other empire has to somehow match it, or risk being conquered. Each time someone develops some new and more powerful technological means of control, the other culture must match or exceed it.

This brings me to ask again: who's in charge?

This is one of the ways the technics themselves control the society.

※ ※ ※

Let's discuss electricity, and through that discussion, look at one of the ways authoritarian technics can destroy our ability to imagine.

One of the (many) ways this culture is killing the planet is through a lack of imagination. I think about this all the time, but I especially thought about this in the wake of the Fukushima nuclear catastrophe, and especially in light of three pretty typical responses I read soon after, each showing less imagination than the one before.

The first came from global warming activist George Monbiot (who normally writes much better stuff), who, just ten days after the earthquake and tsunami, wrote in the *Guardian*, "As a result of the disaster at Fukushima, I am no longer nuclear-neutral. I now support the technology." His position was that the catastrophe—the mass release of highly toxic radiation—was caused not by the routine production and concentration of highly radioactive materials, but rather by a natural disaster combined with "a legacy of poor design and corner-cutting." If the Technocrats can just design this monstrous process better, he seems to believe, they can continue to produce and concentrate highly radioactive materials without causing more accidents. Similar arguments were made after Oak Ridge, Windscale, Three Mile Island, and Chernobyl. And of course similar arguments are made *every time* any authoritarian technics leads to disaster, such as Bhopal, Valdez, and Deepwater Horizon. And of course each time we swallow it anew. You'd think by now we'd all know better. And you'd think it wouldn't take a lot of imagination to see how routinely performing an action as stupendously dangerous as the intentional concentration of highly toxic and radioactive materials would render their eventual catastrophic release not so much an accident as an inevitability, with the question of *if* quickly giving way to the questions of *when, how often*, and *how bad*.

I think the reference we're looking for here is *The Sorcerer's Apprentice*.

The second comment I read came from someone who did not have George Monbiot's advantage of living half a world away from the radioactive mess. In late March of that year, an official with the Japanese nuclear regulatory agency told the *Wall Street Journal* that Japan is not reconsidering nuclear energy in the wake of Fukushima, because "Japan couldn't go forward without nuclear power in order to meet its demand for energy today." He said that a significant reduction in nuclear power would result in blackouts, then added, "I don't think anyone could imagine life without electricity." There's nothing surprising about his response. Most exploiters cannot imagine life without the benefits of their exploitation, and, perhaps more importantly, cannot imagine that anyone else could imagine going through life being any less exploitative than they are. Many slave owners cannot imagine life without slave labor. Many pimps cannot imagine life without prostituting women. Many abusers cannot imagine life without those they routinely abuse. And many addicts cannot imagine life without their addictions, whether to heroin, crack, television, the internet, entitlement, power, economic growth, technological escalation, electricity, or industrial civilization.

The failure of imagination at work here is stunning, or at least it would be had we not already rendered ourselves relatively insensate by our addiction or enslavement to these authoritarian technics, these technics that have become some of this culture's assumptions which must never be questioned. Humans have lived without industrially-generated electricity for nearly all of our existence; we thrived on every continent except Antarctica. And for nearly all those years, the majority of humans lived sustainably and comfortably. And let's not forget the many traditional Indigenous peoples (plus another almost 2 billion people) who are living without electricity today. The Japanese official is so lacking in imagination that he can't even imagine that they exist.

George Monbiot, in his *Guardian* article, asks some questions about living without industrial electricity: "How do we drive our textile mills, brick kilns, blast furnaces and electric railways—not to mention advanced industrial processes? Rooftop solar panels?"

These rhetorical questions are problematical for multiple reasons. The first is that he explicitly identifies with those processes, technics, and people who are killing the planet, and not the real world. How differently would we react to his rhetorical questions if we changed just a few words? "How do *the capitalists* drive *their* textile mills, brick kilns, blast furnaces and electric railways—not to mention advanced industrial processes. Rooftop solar panels?"

The answer? Not our problem. And unless the capitalists can come up with a way to perform these actions without harming other communities, including nonhuman communities, then the real problem we face is: how do we stop them?

Once you break your identification with the system, with the authoritarian technics that are driving planetary murder, your language and your actions become very different. Once you identify with the real, living planet, everything changes.

To be clear: it's not my responsibility to figure out how to deliver the energy that the capitalists "need" to run their factories (no, they *need* to breathe clean air, and drink clean water, and eat nourishing food; they don't *need* to run a factory). Nor is it the responsibility of others who are harmed by their electricity-generation. And if that electricity can't be generated without harming other communities, it shouldn't be generated.

In any case, Monbiot's (one hopes, temporary) identification with the capitalists leads him to a conclusion that makes no sense to someone who is not in thrall to the technics, but that is easily understood once we realize he is being guided not by life, but by the technics itself. He states, "The moment you consider the demands of the whole economy is the moment at which you fall out of love with local energy production." Actually, no. The moment you consider the demands of the whole economy is the moment you fall out of love with the whole economy, an economy that is systematically exploitative and destructive, an economy that is killing the planet.

It is insane to favor textile mills, brick kilns, blast furnaces, electric railways, and advanced industrial processes over a living planet. Our ability to imagine is so impoverished that we cannot even imagine what is happening right in front of our faces.

Why is it unimaginable, unthinkable, or absurd to talk about getting rid of industrial electricity, but it is not unimaginable, unthinkable, and absurd to think about extirpating great apes, great cats, salmon, passenger pigeons, Eskimo curlews, short-nosed sea snakes, coral reef communities, entire oceans? And why is it just as accepted to allow the extinction of Indigenous humans who are also inevitably victims of this way of life (many of whom live with little or no electricity)? This failure of imagination is not only insane, it is profoundly immoral.

Imagine for a moment that we weren't suffering from this lack of imagination. Imagine a public official saying not that he cannot imagine living without electricity, but that he cannot imagine living with it, that what he can't imagine living without are polar bears, the mother swimming hundreds of miles next to her child, and, when the child tires, hundreds of miles more with the cub on her back. Imagine if public officials—or better, imagine if we *all*—were to say we cannot imagine living without rockhopper penguins (as I write this, the largest nesting grounds of endangered rockhoppers are threatened by an oil spill). Imagine if we were to say we cannot imagine living without the heart-stopping flutters and swoops and dives of bats, and we cannot imagine living without hearing frog song in spring. Imagine if we were to say that we cannot live without the solemn grace of newts, and the cheerful flight of bumblebees (some areas of China are so polluted that all pollinators are dead, which means most flowering plants are effectively dead, which means hundreds of millions of years of evolution have been destroyed). Imagine if it were not this destructive culture—and its textile mills, brick kilns, electric railways, and advanced industrial processes—that we could not imagine living without, but rather the real, physical world.

How would we act, and react, differently if we not only said these things but meant them? How would we act, and react, differently if we were not insane? And I mean that in the deepest sense, of being out of touch with physical reality. How can it be so difficult to understand that humans can survive (and have survived) quite well without an industrial economy, but an industrial economy—and in fact any economy—cannot survive without a living planet?

The truth is, the Japanese official and anybody else who states that they cannot imagine living without electricity had better start, because the industrial generation of electricity is simply not sustainable—whether it's by coal or oil or hydro or industrial solar and wind—which means someday, and likely someday soon, people will be not only imagining living without electricity, but actually living without it, along with the more than 2 billion already doing so. About this prospect, a hapa (half Hawaiian) man recently said to me, "A lot of us are just biding our time, waiting to go back to the old ways. Can't be more than a few decades at the latest. We did okay out here without microwave popcorn and weed-whackers and Jet Skis."

Which leads me to the third article I read, titled "What Are You Willing to Sacrifice to Give Up Nuclear Energy?" In it, the author talks, as did the Japanese official, as does more or less everyone for whom this culture's economy is more important than life on the planet, about the importance of cheap energy to the industrial economy. But he's got it all wrong. The real question is: what are you willing to sacrifice to allow the continuation of nuclear energy? And more broadly: what are you willing to sacrifice to allow the continuation of this industrialized way of life?

Given that industrial-scale electricity is unsustainable, and that a lot of people, including nonhuman people, are dying because of it, another question worth asking is: what will be left of the world when the electricity goes off? Just as with the temporary ability of industrial humans to move very fast, we all need to ask for how long will (some) humans have industrially-generated electricity, and at what cost? I can't speak for you, but I'd rather be living on a planet that is healthier and more capable of sustaining life, than on one that is less. And I'm sure nonhumans would as well.

＊ ＊ ＊

Remind me yet again, who's in charge?

＊ ＊ ＊

Lewis Mumford wrote, "My thesis, to put it bluntly, is that from late Neolithic times in the Near East, right down to our own day, two technologies have recurrently existed side by side: one authoritarian, the other democratic, the first system-centered, immensely powerful, but inherently unstable, the other [hu]man-centered, relatively weak, but resourceful and durable. If I am right, we are now rapidly approaching a point at which, unless we radically alter our present course, our surviving democratic technics will be completely suppressed or supplanted, so that every residual autonomy will be wiped out, or will be permitted only as a playful device of government, like national balloting for already chosen leaders in totalitarian countries."[110]

＊ ＊ ＊

Would you like to vote for a Democrat, or a Republican? Would you like a cherry-flavored pellet, or a banana-flavored pellet? Sure, we've all got choices.

Just not the choice to live on a living planet.

＊ ＊ ＊

Mumford wrote that while "democratic technics goes back to the earliest use of tools, authoritarian technics is a much more recent achievement: it begins around the fourth millennium BC in a new configuration of technical invention, scientific observation, and centralized political control that gave rise to the peculiar mode of life we may now identify, without eulogy, as civilization. Under the new institution of kingship, activities that had been scattered, diversified, cut to the human measure, were united on a monumental scale into an entirely new kind of theological-technological mass organization. In the person of an absolute ruler, whose word was law, cosmic powers came down to earth, mobilizing and unifying the efforts of thousands of men, hitherto all-too-autonomous and too decentralized to act voluntarily in unison for purposes that lay beyond the village horizon. The new authoritarian technology was not limited by village custom or human sentiment: its herculean feats of mechanical organization rested on ruthless physical coercion, forced labor and slavery, which brought into existence [social] machines that were capable of exerting thousands of horsepower centuries before horses were harnessed or wheels invented. This centralized technics drew on inventions and scientific discoveries of a high order: the written record, mathematics and astronomy, irrigation and canalization: above all, it created complex human machines composed of specialized, standardized, replaceable, interdependent parts—the work army, the military army, the bureaucracy. These work armies and military armies raised the ceiling of human achievement: the first in mass construction, the second in mass destruction, both on a scale hitherto inconceivable [I would say that both are destructive: the latter is the army for the war against humans, and the former the army for the war against nonhumans]. Despite its constant drive to destruction, this totalitarian technics was tolerated, perhaps even welcomed, in home territory, for it created the first economy of controlled abundance: notably, immense food crops that not merely supported a big urban population but released a large trained minority for purely religious, scientific, bureaucratic, or military activity. But the efficiency of the system was impaired by weaknesses that were never overcome until our own day."[111]

❊ ❊ ❊

Do you remember the story of the chimpanzees who outperform humans at games that require players to perceive and respond to patterns in other players' game play? And do you remember the conclusions the human supremacists reached regarding the chimpanzees' superiority at this game? One was that the nonhumans were using a "simpler model" while humans were "overthinking" it. Another was that chimpanzees are deceitful, manipulative cheaters, and humans, on the other hand, the clearly superior ones, have developed language, semantic thought, and cooperation.

At the time, I made snarky comments about how the people claiming humans are cooperative are among those systematically imprisoning, exploiting, and/or exterminating nonhumans the world over, refusing to participate in (or cooperate with) natural communities, and in their failure to cooperate with other members of natural communities, and instead in their attempts to dominate these communities, are killing the planet. I cannot imagine anything less cooperative than trying to convert the entire planet to use by you and others like you.

The use of the word *cooperation* stuck with me. How could they say something so completely counterfactual and just plain stupid? Yes, I know that believing is seeing, such that your ideology can pretty much determine what you perceive and what you don't. And yes, I understand that human supremacism causes its adherents to project "all things bad" onto nonhumans (and onto the body) and "all things good" onto the wonderful amazing human mind. And yes, I understand that human supremacists believe the human brain is the most complex phenomenon in the universe, and yes, I understand that for human supremacists, all meaning comes only from humans. I saw another example of this latter just yesterday (actually, I see examples of this *all the time*, but I'll share this one). I was reading a G.K. Chesterton *Father Brown* story, and he had the following throwaway paragraph that illustrates yet again this culture's unquestioned belief that humans are the only bearers of meaning: "Far as the eye could see, farther and farther as they mounted the slope, were seas beyond seas of pines, now all aslope one way under the wind. And that universal gesture seemed as vain as it was vast, as vain as if that wind were whistling about some unpeopled and purposeless planet. Through all that infinite growth of grey-blue forests sang, shrill and high, that ancient sorrow that is in the heart of all heathen things. One could fancy that the voices from the underworld of unfathomable foliage were cries of the lost and wandering pagan gods: gods who had gone roaming in that irrational forest, and who will never find their way back to heaven."[112] If there are no "people" there is no purpose. There is no rationality. There is, however, sorrow. And of course, there is no heaven.

Pretty much everything that is wrong with how this culture perceives the natural world in just four sentences.

Anyway, I understand all of this, but still couldn't wrap my mind around the presumption that humans had "developed" cooperation, and most especially that we had done so after we "left the trees." These people have never heard of flowers and bees cooperating in pollination? Salmon and forests cooperating? Hell, bacteria in our own guts cooperating so we can, you know, digest? And I just don't see how a culture that created capitalism, the selfish gene theory, and more broadly, human suprema-cism, and that is destroying the planet, could be even remotely accused of "cooperating." I just read that in the North Atlantic, cod populations are at about 2 to 3 percent of what they once were, and are not recov-ering, but continuing to decline. Yet commercial fishing corporations are refusing to allow measures—such as halting bottom trawling, sufficiently lowering (or eliminating entirely) the catch, and so on—to let the cod have even a chance at recovery. They call this cooperating?

But now I get it. And Mumford helped me understand. Let's take this step by step. Within this supremacist culture our epistemology—how we know whether something is true—is tied to domination. As Dawkins stated, "Science bases its claims to truth on its spectacular ability to make matter and energy jump through hoops on command, and to predict what will happen and when." This tie between domination and episte-mology is generally accepted without question or thought in this culture. Likewise, within this supremacist culture it is authoritarian technics that are generally considered the greatest achievements. This tie between dom-ination and achievement is also generally accepted without question or thought in this culture. So why should it surprise us when the notion of cooperation is likewise coopted into the service of authoritarianism and domination? *Cooperation*, in this supremacist perspective, does not in fact mean reading the needs of those in your community and responding to them by helping these others. And it certainly doesn't mean reading the needs of those in your larger biotic community and acting to improve the capacity of this biotic community to support life. It does not mean cooperating with the living planet to make this living planet healthier, as is normal behavior for residents of this planet.[113]

No, *cooperation* in this context means something completely different, something completely in line with the thrust of this whole authoritarian culture. *Cooperation* in this case means the creation of "complex human machines composed of specialized, standardized, replaceable, interdependent parts—the work army, the military army, the bureaucracy." *Cooperation* has been perverted into its toxic mimic through the conversion of living human animals into cogs in hierarchical social machines.

❀ ❀ ❀

In the case of wiping out the cod, *cooperating* means forming corporations to control armies of workers who are "cooperating" to build huge fishing vessels; forming academic bureaucracies to task armies of researchers to "cooperate" to discover ways to use sonar to find and destroy schools of cod; forming corporations to control armies of fishermen to "cooperate" in killing fish; forming corporations to charge armies of workers with "cooperating" to transport fish to markets; forming corporations to send armies of lobbyists to "work together" with "decision-makers" to make sure the catch doesn't go down so long as there is a single cod who can be turned into fish sticks (and hence, money); and forming huge armies of "fisheries scientists" and bureaucrats to cooperate in overseeing the extermination of the cod, via managing them to death.

❀ ❀ ❀

Gosh, it's a lot more flattering to say that humans are superior because we learned to "cooperate," rather than to say we're superior because we learned the power of top-down, military-style bureaucratic organization, isn't it? Although this organizational form *does* bring a lot of benefits (that is, for the few at the expense of the many, including nonhumans); and it's also completely fantastic at getting large numbers of perhaps otherwise moral people to act in profoundly immoral ways.

For the sake of our own vanity and sense of superiority, let's keep calling it "cooperation," okay?

❀ ❀ ❀

Mumford then describes some of the weaknesses of the authoritarian system: "To begin with, the democratic economy of the agricultural village resisted incorporation into the new authoritarian system. So even the Roman Empire found it expedient, once resistance was broken and taxes were collected, to consent to a large degree of local autonomy in religion and government. Moreover, as long as agriculture absorbed the labor of some 90 per cent of the population, mass technics were confined largely to the populous urban centers. Since authoritarian technics first took form in an age when metals were scarce and human raw material, captured in war, was easily convertible into machines, its directors never bothered to invent inorganic mechanical substitutes. But there were even greater weaknesses: the system had no inner coherence: a break in communication, a missing link in the chain of command, and the great human machines fell apart. Finally, the myths upon which the whole system was based—particularly the essential myth of kingship—were irrational, with their paranoid suspicions and animosities and their paranoid claims to unconditional obedience and absolute power. For all its redoubtable constructive achievements, authoritarian technics expressed a deep hostility to life."[114]

As we see.

CHAPTER FOURTEEN

THE DIVINE RIGHT OF MACHINES

The State of monarchy is the supreme thing on Earth. . . . As to dispute what God may do is blasphemy, so is it treason in subjects to dispute what a king may do.

KING JAMES I

Years ago my friend Frances Moore Lappé told me she derives a certain amount of optimism from the question, "Why did people stop believing in the Divine Right of Kings?" Her answer? "They just did. At one point they believed that kings were put on the throne by God, and then at some point they didn't. My optimism comes from the fact that they just stopped believing in this destructive notion. We can do that with other destructive notions as well."

I wish I shared her optimism. But it seems clear to me that people have not, in fact, stopped believing in the Divine Right of Kings. This belief, and the Great Chain of Being that rationalizes it, runs our culture more now than ever before; it is just that the insanity, megalomania, power-lust, feelings of specialness and superiority, and claims to unbounded power that used to be associated specifically with royalty have now spread to the widest reaches and most formative depths of this supremacist culture. The Divine Right of Kings has not been abandoned. It has morphed into the Divine Right of Humans, especially the Divine Right of Industrial ("Developed") Humans. Even worse, it has morphed into the Divine Right of Machines.

❊ ❊ ❊

Mumford again: "That authoritarian technics has come back today in

231

an immensely magnified and adroitly perfected form. Up to now, following the optimistic premises of nineteenth century thinkers like Auguste Comte and Herbert Spencer, we have regarded the spread of experimental science and mechanical invention as the soundest guarantee of a peaceful, productive, above all democratic, industrial society. Many have even comfortably supposed that the revolt against arbitrary political power in the seventeenth century was causally connected with the industrial revolution that accompanied it. But what we have interpreted as the new freedom now turns out to be a much more sophisticated version of the old slavery: for the rise of political democracy during the last few centuries has been increasingly nullified by the successful resurrection of a centralized authoritarian technics—a technics that had in fact for long lapsed in many parts of the world. Let us fool ourselves no longer. At the very moment Western nations threw off the ancient regime of absolute government, operating under a once-divine king, they were restoring this same system in a far more effective form in their technology, reintroducing coercions of a military character no less strict in the organization of a factory than in that of the new drilled, uniformed, and regimented army. During the transitional stages of the last two centuries, the ultimate tendency of this system might be in doubt, for in many areas there were strong democratic reactions; but with the knitting together of a scientific ideology, itself liberated from theological restrictions or humanistic purposes, authoritarian technics found an instrument at hand that has now given it absolute command of physical energies of cosmic dimensions. The inventors of nuclear bombs, space rockets, and computers are the pyramid builders of our own age: psychologically inflated by a similar myth of unqualified power, boasting through their science of their increasing omnipotence, if not omniscience, moved by obsessions and compulsions no less irrational than those of earlier absolute systems: particularly the notion that the system itself must be expanded, at whatever eventual cost to life."[115]

❊ ❊ ❊

Actually, things are much worse than either Mumford or I have so far presented.

It should be clear by now that authoritarian technics run the society, such that the culture as a collective cannot imagine living without, for example, industrially-generated electricity; and such that when faced with the murder—sorry, reorganization—of the planet through global warming, this culture's response is to continue the burning of oil, gas, and coal; and to continue constructing dams, cutting down forests for "biomass,' and constructing industrial wind and solar, all of which harm the planet. Those who care could keep adding to this list of authoritarian and destructive technics that rule this society until the world is dead. And even up to the last moment, most people won't care, so long as they can somehow still rationalize their own feelings of superiority.

※ ※ ※

Noam Chomsky and others, for example Ernst Mayr, have argued that intelligent life is not sustainable. In an essay entitled "Human Intelligence and the Environment," Chomsky writes that Mayr "basically argued that intelligence is a kind of lethal mutation. And he had a good argument. He pointed out that if you take a look at biological success, which is essentially measured by how many of us are there, the organisms that do quite well are those that mutate very quickly, like bacteria, or those that are stuck in a fixed ecological niche, like beetles. But as you go up the scale of what we call intelligence, they are less and less successful. By the time you get to mammals, there are very few of them as compared with, say, insects. By the time you get to humans, the origin of humans may be 100,000 years ago, there is a very small group. We are kind of misled now because there are a lot of humans around, but that's a matter of a few thousand years, which is meaningless from an evolutionary point of view. His argument was, you're just not going to find intelligent life elsewhere, and you probably won't find it here for very long either because it's just a lethal mutation. He also added, a little bit ominously, that the average life span of a species, of the billions that have existed, is about 100,000 years, which is roughly the length of time that modern humans have existed.

"With the environmental crisis, we're now in a situation where we can decide whether Mayr was right or not. If nothing significant is done about it, and pretty quickly, then he will have been correct: human intelligence is indeed a lethal mutation. Maybe some humans will survive, but it will be scattered and nothing like a decent existence, and we'll take a lot of the rest of the living world along with us."[116]

Chomsky is a brilliant thinker and writer. He has done more than almost any other person in the last fifty years to expose United States imperialism. But these statements—and this is true for many of Chomsky's comments on the natural world—reveal how decisively unquestioned beliefs in human supremacism affect discourse.

We shouldn't be surprised that Mayr and so many others believe intelligence is lethal. This culture teaches us that the way we know something is true is by controlling others: by forcing matter and energy to jump through hoops on command. The dominant cultural narrative tells us that our greatest achievements are those that facilitate our domination of others. This culture conflates "cooperation" with top-down organizational systems that have as their function the multiplication of power. Is it any wonder, then, that members of this culture believe that intelligence is "lethal"?

But how can we stop the murder of the planet if leading intellectuals label intelligence itself as "lethal," and say that the murder of the planet is a result of this intelligence?

I want to deconstruct a few of these comments before I get to the main point of bringing this up. In the essay I've been quoting from, Chomsky writes, "If you take a look at biological success, which is essentially measured by how many of us are there . . ." But this is not an appropriate or realistic measure of biological success. Instead it is one that is based on this culture's model of overshoot and conquest. We've all been taught that life is somehow like a computer game, where your success is measured by how many points you rack up; or like *Risk*, where your success is measured by how many little plastic armies you have and how much of the map you control. But this measure of biological success is simply the same old Biblical commandment to go forth and multiply projected onto the natural world. It's also, since they come from the same imperative, a projection of the dominant economic mindset onto the natural world, a capitalist mindset where your success is measured by how many dollars or how many franchises you own. Switching terms again, but still coming from the same imperative, it's a projection onto the natural world of a colonialist or imperialist mindset where your success is measured by how much of other people's land you take over for your own use and to increase your numbers. That's the definition of a colonialist mindset. And it is precisely how this culture has maximized its numbers—succeeded, according to this metric—by taking over someone else's land (in this case, land needed by both Indigenous humans, and nonhumans).

So many anti-imperialists understand all this when it comes to economic and social policy, but it is a measure of the hold that human supremacism has over our minds and our discourse that these same anti-imperialists—and, in fact, most of us—cannot see that the definition of biological success they use is precisely the measure of success for colonialism or empire. In this case it is simply human empire, or more precisely, an empire of authoritarian technics.

I would argue that a far better measure of biological success would be whether the presence and population of a given species improves the health and resilience of the larger biotic community of whom it is a member and on whom it relies for sustenance, thereby ensuring its own species' survival as well as the survival of other members of its biotic community. How would we act differently if we allowed this definition of ecological success to influence our social policies?

Next, please note the phrase "up the scale." That is directly from the Great Chain of Being. But there is no Great Chain of Being that goes from unintelligent nonhumans to lethally intelligent humans.

Now to the point, which is that I'm really uncomfortable with intelligence being labeled as lethal, mainly because I don't think it's accurate, but also because it naturalizes the destructiveness. This is why, as every anti-imperialist knows, colonizers nearly always attempt to justify as right or natural their status at the top of the hierarchy they themselves created. (It's funny, isn't it, that the ones who create, then articulate these "natural" hierarchies so often end up at the top. What are the odds?) I'm also uncomfortable because it doesn't make any sense to me that even when we do literally the stupidest thing possible, which is to kill the planet that is our only home and that supports our lives, it is a sign of our intelligence. We're so damn smart that we're maladaptive.

But this is one of the ways supremacisms control our thought and discourse: no matter what evidence is presented, even if it is damning to the supremacist's in-group, we'll find *some way* to use it to reinforce our sense of superiority.

* * *

Here's another thing that kills me about the notion that intelligence is lethal, or that humans are killing the planet because we're so damn smart. At the same time that *some* humans have been killing the planet, and through many of the same processes, and for many of the same reasons, those same humans who have been killing the planet—the civilized; those who are enslaved to authoritarian technics—have been killing Indigenous humans, overrunning, committing genocide against, and often exterminating them. If you were to look at a time-lapse map of worsening ecological conditions and superimpose upon it a time-lapse map of the expansion of civilized, agricultural peoples, and over that a time-lapse map of land stolen from Indigenous peoples, you would find that the maps were pretty much the same. Yet somehow, public intellectuals—and a lot of them—can get by saying that the destruction is caused because humans as a species are so damn smart. And a lot of listeners in this human supremacist culture nod their heads and thoughtfully rub their chins, NPR-style. But the same processes that led, and lead, to the murder of the planet also led, and lead, to land theft from Indigenous humans (in the former case it's land theft from nonhumans, and in the latter it's land theft from both the humans and the nonhumans with whom they live). How would these same listeners respond if the public intellectuals said that *some* humans, by which they meant civilized humans (including, ahem, whites), have been able to overrun Indigenous nations (including those made up of, ahem, people of color) because of the superior intelligence of the conquerors? "Oh, the Europeans conquered North America and destroyed hundreds of Indigenous cultures because the Europeans are far more intelligent, and intelligence is a lethal mutation." How would that sound? Because that's really what they're saying.

* * *

Ah, but what if they have a point? What if "intelligence," as defined by human supremacists, *is* lethal? I'm not saying that civilized humans are smarter than Indigenous humans (or, for that matter, anyone else on the planet). I'm saying, what if the primary form of intelligence we *recognize*, we reward, we encourage, we worship; what if *that* form of intelligence is lethal?

❀ ❀ ❀

There are at least two other problems with blaming "intelligence" for the murder of the planet. The first—and this is, of course, one of the reasons it's done—is that it fuels this already-supremely-narcissistic culture's narcissism. This is the worst thing you can do with a narcissist. "Oh, we're killing the planet because we're so smart? How kind of you to say that. Of course it's nothing less than what we deserve. . . . We're completely fabulous, don't you agree? Even the oceans agree. And if they don't, we'll kill them. The planet just cannot handle our raw intelligence."

The second reason is that saying our "intelligence" is a "lethal mutation" transforms the murder of the planet from the ongoing result of lots of very bad and very immoral ongoing social choices, which we can name, and which provide benefits for some classes of people at the expense of others, into something beyond our control, into something we can do nothing about, into a classic tragedy, with us starring (of course) as the tragic hero whose tragic flaw is that he is just too damn smart for this world.

I'm sure that would play well to the right audience.

I'm not sure Indigenous humans or nonhumans, though, would like it very much.

The notion of being so smart that we kill the planet is pretty much the ultimate oxymoron (emphasis on moron). I know I've said this before, but I'm going to keep saying it because we as a culture are clearly collectively not getting it: there is no action any species could take that would be more completely, fundamentally, unforgivably stupid than to harm the capacity of the planet to support life. The planet that is our only home.

It takes world-class stupidity to foul the entire planet.

And that's really the point here. By calling the murder of the planet an act of intelligence, one is encouraging that destructiveness. Smart is good, right? We'd rather be smart than not smart, right?

How would our society as a whole act differently if, instead of portraying the acts of destroying forests or killing oceans as signs and validations of our intelligence, we were to speak honestly about them, and say that they are acts of mind-boggling stupidity? How would we act differently if public intellectuals argued that this culture is killing the planet because we're so fucking stupid? Wouldn't that change our behavior?

Of course if someone were to argue that humans are killing the planet because humans are lethally stupid, I would still point out that plenty of Indigenous cultures did not destroy their landbases. So I would argue that it is not that humans are stupid, but that this culture makes people stupid, in fact so stupid that they would rather kill the planet that is the source of our lives and the lives of all these other beautiful beings with whom we share this planet, than to acknowledge that they are making stupid social choice after stupid social choice.

❄ ❄ ❄

And how do we look at other atrocities? We don't wring our hands and say that the Nazis committed genocide because they were too damn smart. The Catholic Church didn't promulgate the Inquisition because of the intelligence of Popes Innocent or Gregory or Lucius. Whites didn't enslave Africans because whites are smarter. Rapists don't commit rape because they're smarter than their victims. Why can't we just acknowledge that atrocities are atrocities? Sure, every group that commits atrocities has already built up a philosophy to justify these atrocities. But atrocities are not tragedies, and the perpetrators of atrocities are not tragic heroes. They're just people who, for this reason or that, have talked themselves into rationalizing and then committing atrocities, and then rationalizing them again and again as they continue to commit them.

Let's not make the committing of atrocities into something it isn't. With all the world at stake, let's at least be that honest.

CHAPTER FIFTEEN

AGRICULTURE

The adoption of agriculture, supposedly our most decisive step toward a better life, was in many ways a catastrophe from which we have never recovered. With agriculture came the gross social and sexual inequality, the disease and despotism [and the ecological destruction, and militarism], that curse our existence.

JARED DIAMOND

Noam Chomsky, who is, again, one of the most important public intellectuals of the late twentieth century, also says, about agriculture and energy, "If agriculture is inherently destructive, we might as well say good-bye to each other, because whatever we eat, it's coming from agriculture, whether it's meat or anything else, milk, whatever it is. There is no particular reason to believe that it's inherently destructive. We do happen to have destructive forms of agriculture: high-energy inputs, high fertilizer inputs. . . . So are there other ways of developing agricultural systems which will be basically sustainable? It's kind of like energy. There's no known inherent reason why that's impossible."

Once again, he's not alone. He has an entire culture for company. At this point, nearly all writers and historians and scientists share this worldview, even those who are revolutionary and/or radical in other ways. It's depressing as hell.

I guess my question would be, if the entire history of agriculture—six thousand years of destroying every biome it has touched—doesn't constitute "reason to believe" that agriculture is inherently destructive, what, precisely, would constitute evidence? What will be our threshold to finally acknowledge this? Seven thousand years? Eight thousand? The complete destruction of the biosphere? I doubt if even those will suffice.

Here's a particular reason to believe in agriculture's destructiveness: black-skinned, pink-tusked elephants in China.

You've never heard of these? That might be because they were exterminated by agriculture. Not modern agriculture. Agriculture.

Here's another reason: Mesopotamian elephants.

You've never heard of these either? That might be because they, too, were exterminated by agriculture. Not modern agriculture. Agriculture.

Carolina parakeets. Prairie dogs. Bison. The (formerly) Fertile Crescent. Iowa, which was once one of the most biologically diverse places in North America. The Everglades. Monarch butterflies. All devastated by and for agriculture.

Dead zones in oceans.

The Mississippi River. The Colorado River. Every river on the planet who has significant agriculture within its watershed.

Every Indigenous nation on the planet decimated or exterminated by the conquest that necessarily accompanies agriculture and consequent overshoot and conquest.

How much evidence do we need?

Agriculture destroys more nonhuman habitat than any other human activity. This has been true from the beginnings of agriculture. This destruction of habitat is not a by-product of agriculture. It is the *point* of agriculture: to convert land specifically to human use, and then to impede succession, that is, to stop the land's attempts to heal itself. And the fact that the central acts of agriculture—destroying habitat and disallowing it from healing—are harmful to the natural world is not a reason to believe that agriculture is necessarily destructive?

Agriculture destroys soil, the basis of terrestrial life. One way it kills soil is through causing erosion. It is the leading anthropogenic cause of erosion. What do you expect to happen when you remove from soil the protective covering of plants? The plants were there for a reason (oh, that's right, I forgot, there is no purpose or reason or function or intelligence in nature). Removing groundcover, which is the function of the plow, one of this culture's greatest achievements (and yes, plows have function while plants don't), is the equivalent of flaying the biome who is being converted to cropland. Would anyone say there's no reason to believe that flaying someone harms them?

Agriculture destroys water quality. Erosion hurts not only the land being eroded, but the waters choked by more sediment than they need or want. Of course irrigated farmland takes water, water that was, until it was removed from the river, lake, or ground, someone else's home. Primarily because of agriculture, a full 25 percent of the world's rivers no longer reach the ocean or sea. This includes such once-huge rivers as the Colorado, the Indus, the Amu Darya, the Syr Darya, the Rio Grande, the Yellow, the Teesta, the Murray, and so many others. The dewatering of these rivers destroys the rivers, the wetlands, the estuaries, the seas and oceans who need these waters. What percent of the world's rivers would have to be murdered before we can consider this evidence of agriculture's destructiveness? Forty percent? Sixty? All of them?

Right now 115 percent of the water in the Colorado River is allocated for "beneficial use," primarily agriculture. Yes, 115 percent. And governments are building more pipelines to take yet more water.

Would allocating more water from a river than the river carries, and then building more pipelines to carry away even more water, be considered a sign of intelligence?

Because of agricultural runoff, there are more than 450 dead zones in the oceans. Dead zones. Devoid of aerobic life. How many must there be before we can conclude that agriculture is inherently destructive?

And do we really need to talk about the unsustainability of so-called pesticides, and the relative intelligence of putting poison on your own food, in fact covering the planet in poisons? And pesticides raise another issue about the inherent unsustainability of agriculture. In order to make their food species easier to control, agriculturalists reduce or eliminate the natural defenses of the target species. This has often been done through breeding programs, selectively breeding for docile plants who devote their energies to making themselves nutritious and tasty instead of making themselves toxic or unpalatable, and selectively breeding for docile animals who are more likely to put on meat than to fight or run. But now, since these plants and animals have fewer defenses, the agriculturalists step in to kill those who would eat these now-relatively-defenseless plants and animals.

We could talk extensively about the tremendous harm all of this has done and must do to the real world—including the toxification of the total environment—but my point here is that all of this is not only functionally unsustainable on a physical level, but it creates an unsustainable mindset, for many reasons. One is that because we have formed or deformed these other beings—plants, animals, and others—to suit our needs, we can come to believe that we and not the planet are their creators. We can come to believe that we and not the planet are the creators of life. We can come to believe that we are smarter than we are (and smarter than they are, no matter who "they" are). Because we can "make" Cornish X chickens—it's actually the hens who make the babies, but you know what I mean—with so much breast meat they suffocate if you allow them to grow up; and because we can "make" dogs—same caveat applies—small enough to fit in a person's pocket, this then causes too many of us to believe we can (and should) make the entire world do what we want. We can make it jump through hoops on command. In the words of Charles Mann, "Anything goes."

Another is that because these domesticated plants and animals are reliant upon us for their survival, and because keeping these scarred and domesticated landscapes scarred and domesticated requires we constantly fight to keep nature from reclaiming (i.e., healing) these lands, we can come to believe that our position as wards of the pig sty—that is, stywards—qualifies us to be "stewards" of the entire planet. I know this sounds like a ridiculous jump—from being kind of in charge of a small piece of land and a few animals (and importing resources and externalizing harm to do it); to trying to manage (and steal from) incomprehensibly complex webs of interspecies communal relationships without terminally fucking them up—but more or less all of this human supremacist and nature-hating culture has made it, from the right-wing Christians and capitalists who believe we're supposed to subdue the planet, to the lefty Christians and capitalists who believe we're supposed to be stewards, to the loggers whose mud-splattered 4x4s sport the bumper sticker "Healthy forests need loggers," to the anti-environmentalists-pretending-to-be-environmentalists who believe the entire world is a rambunctious garden for us to control. As human supremacist and anti-environmental author Emma Marris puts it, "We are already running the whole Earth, whether we admit it or not." Emma, the word you are looking for is not *running*, it's *ruining*. This insane—and maybe just the teensiest bit narcissistic—belief that the entire world is ours to "run" comes to be one of those assumptions we must never question.

And since the natural world is always trying to reclaim (i.e., heal) the land you've converted solely to human use, it can be remarkably easy to start to believe that the source of the necessities of your life is your own creations, and that nature is not only *not* the source of the necessities of your life but rather the *enemy* of all that is the source of the necessities of your life, the enemy of your creations, the enemy of you. This can lead to a perception of the necessity of a perpetual war against nature. As we see.

This is a big problem.

They say that one sign of intelligence is the ability to recognize patterns. This pattern of overshoot, habitat destruction, destruction of topsoils, destruction of rivers, wetlands, estuaries, oceans, destruction of the lifeways of the Indigenous humans and nonhumans who live in these places, has been happening since the beginning of agriculture, and continues not only unabated but accelerated to this day. Yet, still, so many members of the self-declared most intelligent species on the planet fail (or refuse) to see this pattern. Perhaps when there are no elephants, perhaps when there is no topsoil, perhaps when no rivers reach their destinations, perhaps when the oceans are one big dead zone, maybe then a few of us will acknowledge that agriculture is inherently destructive.

Don't count on it. We're not that smart.

❀ ❀ ❀

Or maybe we're just way too smart for the planet's good, lethally smart, so smart that we, like the humans who regularly lose to imprisoned chimpanzees in games that require we pay attention to what someone else is doing, are overthinking it, and we need a simpler model.

❀ ❀ ❀

Nah. It's more likely that we're just not that smart.

❀ ❀ ❀

Emma Marris says we're "running the whole Earth," and she acts as though this is a good thing, or at least a fact of life we need to accept. But look around. Us "running the whole Earth" is killing it. How completely unintelligent would a person have to be—would a culture have to be—to not have learned, after 6,000 years of ruining every place we attempt to run—and this is true of every civilization that has ever existed—that "running the whole Earth" is a complete fucking disaster?

Every single biome on the planet whom human supremacists have tried to manage has been dramatically harmed. Every single one. There has been not a single success, in terms of biotic health. How fucking arrogant, and how fucking stupid, do you have to be to not be able to discern the pattern in this?

Further, and this is the real point, how much intelligence does it take to cut through the rhetoric and see that "agriculture" and "running the whole Earth" are euphemisms for "stealing"? Converting the entire landbase to human use is certainly stealing from the nonhumans who live there, and it is stealing from the humans living with those nonhumans, and it is stealing from those who would have lived in the future. "Running the whole Earth," likewise, is just another way to say "taking whatever we want and fuck everyone else." I hate to break the news, but while using a euphemism may salve your conscience, and in the case of "running the whole Earth," may make you feel superior, it doesn't change material reality. And I really hate to break the news on this one, too, but material reality is more important than the words we use to describe it. Reality is also more important than our perceptions of it or our beliefs about it.

The only way that anyone can say that agriculture—and other forms of "running the whole Earth"—aren't inherently destructive is by ignoring the costs to all of those who are not in the supremacist class, from the soils to the plants to the oceans to the animals (human and otherwise) to the bacteria to everyone else.

It's just the Doughnut Supreme model all over again.

Human supremacists have one trick, and they do it very well: take from everyone else, and ignore the consequences.

* * *

We can, of course, perform this same exercise for energy usage by this culture, which is functionally just as unsustainable, and so can never be sustainable. We can look at the long and painful history of what energy extraction has done to the planet, and to the Indigenous humans and nonhumans who have been or are unfortunate enough to live near extractable energy. Right now the Maasai, who have been living more or less sustainably in place for at least the last 500 years, are being driven off their land so Kenya can get at the geothermal energy sources beneath.[117] So, and this is something we see time and again, whether it is Indigenous peoples in South America or Asia or India or anywhere else being driven off their land for dams or any other form of industrial energy production, people living sustainably are being driven off the land in the name of "sustainable energy production."

Tell me again, who are the ones who are cooperative? Who are the ones who are intelligent? Who are the ones who are superior?

❋ ❋ ❋

Because we're talking about some of the unquestioned beliefs that are the real authorities of this culture, I need to be as clear as possible. Agriculture—by which I don't mean hunting and gathering, or horticulture, or pastoralism, but agriculture—leads to overshoot. That's what happens when you convert the land solely to human use. Converting land solely to human use is by definition inherently destructive to all others who live there. I am consistently stunned at the number of people who have convinced themselves that taking from others does not harm those from whom one is stealing.

And agriculture is not and can never be sustainable. Permaculturalist and author Toby Hemenway wrote a brilliant essay entitled, "Is Sustainable Agriculture an Oxymoron?" It begins, "Jared Diamond calls it 'the worst mistake in the history of the human race.' Bill Mollison says that it can 'destroy whole landscapes.' Are they describing nuclear energy? Suburbia? Coal mining? No. They are talking about agriculture. The problem is not simply that farming in its current industrial manifestation is destroying topsoil and biodiversity. Agriculture in any form is inherently unsustainable. At its doorstep can also be laid the basis of our culture's split between humans and nature, much disease and poor health, and the origins of dominator hierarchies and the police state."[118]

Further, since overshoot is by definition not sustainable (or else it wouldn't be overshoot), and since humans (especially industrial humans) have already far exceeded the carrying capacity of the earth, there are, once again by definition, no sustainable ways to support this number of (especially industrial) humans. For the third time, that's the *definition* of *overshoot*. In overshoot, sustainability is *by definition* not possible. So even if agriculture had ever been sustainable, that implausible possibility has been foreclosed.

And if you don't think that (especially industrial) humans have overshot carrying capacity, then I'm afraid we don't have a lot to say to each other; I prefer my conversations to be based on physical reality, and if you can't hear the 200 species extirpated per day and the acidifying oceans and forests and wetlands and prairies reduced to less than 2 percent of what they used to be all screaming "overshoot," then you ain't listenin'.

Of course, acknowledging that agriculture is inherently destructive doesn't mean we should kiss each other goodbye. It just means that if we want to stop this culture from killing the rest of the planet, we should start by being honest about the circumstances in which we find ourselves.

That's not so hard, is it?

❋ ❋ ❋

Any form of sustainable food procurement (I don't want to say *food production*, since the Tolowa didn't produce salmon, but caught and ate them) will have to not be human supremacist. Human supremacism is unsustainable.

❋ ❋ ❋

Here's the real reason I brought all this up. Even faced with the murder of the planet, most people who have been inculcated into this culture refuse to question human supremacism and human empire. They would sacrifice life on this entire planet rather than question whether we really are more intelligent than anyone else; they will blame what they call intelligence rather than question the unquestioned beliefs that motivate this culture. They are *already sacrificing* life on the planet rather than questioning their assumptions.

This is one of the ways an authoritarian technics is authoritarian: like any other despot, it cannot be questioned, even when it is killing all we hold dear, all we truly need to survive.

This is what we're up against.

I'll be clear. Imperialism can be defined as the taking (by force, threat of force, or even "persuasion," if the power relations between the parties are grossly unequal) of another's land or other "resources" for use at the center of empire. Using this definition, agriculture is imperialism, both against the land and against (human) people of the land. And if we change a few words in the quote at the beginning of the chapter we could easily hear it coming from the mouth of your standard apologist for empire (which, when it comes to intrahuman empire, Chomsky definitely is not), "If empire is inherently destructive, we might as well say good-bye to each other, because all of our energy and consumer goods come from empire, whether it's coal from the internal colonies of Appalachia and the High Plains, tin from Bolivia, clothes from sweatshops in Haiti or Vietnam, steel from the slave-based factories in Brazil. Whatever it is. There is no reason to believe that empire and colonialism are inherently destructive."

What would any reasonable anti-imperialist say to someone who said these same things? I think the analysis would be similar to what I've done here.

Empire happens for material reasons. German Reichskanzler Paul von Hindenberg described the relationship perfectly: "Without colonies no security regarding the acquisition of raw materials, without raw materials no industry, without industry no adequate standard of living and wealth. Therefore, Germans, do we need colonies."[119] You can't have high speed rail without mines and smelters. You can't have mines and smelters without empire. The fact that our way of life is dependent upon this empire is no reason we should not discuss it.

❈ ❈ ❈

I can't stop thinking about these comments implying that because our way of life is dependent upon agriculture, then somehow we cannot and most importantly must not question agriculture's inherent destructiveness. And I can't stop thinking about how much they remind me of that Supreme Court ruling that if this way of life is based on land theft and genocide, then such land theft and genocide "becomes the law of the land, and cannot be questioned." So, because we have enslaved ourselves to land theft and ecocide through agriculture, then any critique of agriculture must be dismissed by the suggestion that if agriculture is destructive we may as well say goodbye to each other?

No. If agriculture is inherently destructive, we should address this honestly. And if our way of life is based on agriculture, and if agriculture is inherently destructive, that provides all the more urgency to making an honest analysis. It's like my doctor friend says about the first step toward cure being proper diagnosis. Well, if we're going to short-circuit diagnosis before it even starts, then there can never be a cure. We are guaranteeing the continued murder of the planet.

I'm not interested in rationalizing the further murder of the planet. We need to face reality, no matter how painful.

CHAPTER SIXTEEN

FACING REALITY

God sends ten thousands truths, which come about us like birds seeking inlet; but we are shut up to them, and so they bring us nothing, but sit and sing awhile upon the roof, and then fly away.

HENRY WARD BEECHER

I'm known for saying that civilization is killing the planet, and that it needs to be stopped before it kills what or who is left. I don't say this because I hate hot showers or Beethoven's Ninth. I say this because I've long been capable of doing simple math.

I can do subtraction. I know if there are six billion passenger pigeons, and you subtract a billion, and then another billion, and you keep subtracting them faster than they can add to their own population (and faster than they can feed all those others in their biotic communities who eat them), then eventually there will be none. I know if there are uncountable salmon, and you reduce their numbers to where you can count them, and then you keep subtracting, eventually there will be none. I know if you estimate the weight of all the fish in the oceans in 1870, and you call that 100 percent, and then you keep subtracting fish until by 2010 you get it down to 10 percent, then there's something deeply wrong with what you're doing. I know the same is true for native forests reduced from 100 percent to 2 percent, native grasslands and wetlands reduced the same.

And I want to bring down civilization because I know how to add. I know that if you take a number, say, 315 (as in parts per million), and keep adding to it, eventually you'll get to 350. And if you keep adding to that you'll get to 400. And if you keep adding to that you'll get to hell.

I don't understand why so many of us don't seem to know how to subtract or to add. Oh, sure, I understand that people come up with lots of rationalizations for avoiding the simple math, and they come up with lots of fancy names and algorithms to attempt to convince themselves that 100 minus 90 doesn't equal 10, or that 315 plus 85 doesn't equal 400 or that somehow hot showers, Beethoven's Ninth, and high-speed internet access for some of us all add up to more than life on earth, but whether you call it "managing forests," "generating hydroelectric power," "developing natural resources," "sustainable development," "green energy," "agriculture," "running the whole Earth," or any of a thousand other names, the subtraction and the addition continue.

What makes the whole thing even more insane is that the economic system requires constant addition, and this addition requires and creates subtraction, by which I mean capitalism (and before it, civilization) requires that production grow—add 2 or 3 percent each year—and production is a measure of the subtraction, that is, of the conversion of the living into the dead: forests into 2x4s, schools of fish into fish sticks or sushi or fertilizer.

The math is both simple and tragic.

I think that for some people—especially those in power—the only math that matters is constant addition into their bank accounts.

But I think that so many of the rest of us do what we can to avoid this math because if we do the subtraction, do the addition, our own personal sum will be unbearable sorrow, terror, and a feeling of being entirely out of control. I think many of us do what we can to avoid this math because we know that if we do the subtraction, do the addition, our psyches and our consciences and our lives will forever be changed; and we know that no matter how fierce the momentum that leads to this subtraction and addition, no matter our fears that we may be crushed (or perhaps more fearsome, ridiculed), that we will be led in some way to oppose the subtraction of life and the addition of toxics to this planet that is our only home.

I'll tell you my fear, and I'll tell you my dream. My fear is that the subtraction and the addition will continue until there is nothing left on this planet but ashes and dust. That the oceans whom this culture has caused to go from 100 to ten continue down to zero. That the 200 species this culture causes to go extinct each day increases to 300, then 400, then 500, then 600. That the migratory songbird populations who have collapsed, then collapsed, then collapsed, disappear into that eternal night of extinction. That the bumblebees and dragonflies and bats and spiders and sowbugs, whom I already see far less frequently than even a few years ago, disappear as well. That the plastics that now outnumber the phytoplankton by ten to one increase that ratio to 100 to one, 1,000 to one, and so on. That what was 315 and then 350 and then 400 continues to rise. In other words, rationalizations and fancy names and algorithms notwithstanding, that business continues as usual.

I'll tell you another fear: that the subtraction and the addition will last even one more day. Because for this two hundred species—and for all of us—that one day is one day too many.

Of course, business as usual can't continue forever. I understood that when I was a child. We all—by which I mean those of us with any sense whatsoever—know this can't go on; you can't continue to subtract life and add toxics forever. But it has gone on long enough to reduce a number from 100 to ten, to reduce another from 6 billion to zero, another from 100 to two. In other words, it has gone on far too long. And it can reduce those numbers to zero.

I've never forgotten something my dear friend and environmental mentor John Osborn said to me, about why he does his work: "We cannot predict the future. As things become increasingly chaotic I want to make sure some doors remain open." What he means by this is that if lynx and Selkirk caribou and bull trout are alive in ten years, they may be alive in 100. If they're extinct in ten years, they're gone forever. If he can help keep this or that patch of old growth standing for ten years, this or that river free flowing for ten years, it may be alive in 100. If it's cut or dammed now, the damage is done. He's saying, "These will not go down on my watch."

So here is my dream, and what I spend my life working toward: that subtraction and addition switch places, so that each day there is more life on this planet, more fish and birds and insects, more forests and free-flowing rivers and grasslands and wetlands; and fewer toxics. This won't happen because of rationalizations or fancy names or new algorithms. This will only happen because the social conditions—on every level, from the epistemological to the infrastructural—that lead to the subtraction and addition completely change.

That will happen. The only question will be what's left of the planet when it does.

In the meantime, it's up to each of us to ask what we love, and then to defend that beloved. It's time for each of us to say, "Not on my watch."

❁ ❁ ❁

Think about it. You're driving a car down a tunnel at 100 miles per hour directly at a brick wall. Do you turn to the passengers and say goodbye? Do you tell them, "Our lives depend on this car not crashing. I see no evidence of a brick wall. And I see no evidence that car crashes must be inherently destructive"? No, you hit the brakes so hard your foot goes through the floor. If you can stop, great. If you can't, it becomes a question of increasing your odds of survival. I'd rather hit the wall at 90 than 100, 80 than 90, 70 than 80. With your own life and the lives of those you love at stake, every mile per hour you cut away counts.[120]

But what does this culture do? It keeps its foot firmly on the gas.

Now, let's say you're a passenger in this car. What do you do? Do you turn to those in the back seat and say goodbye? Do you pretend there is no brick wall? Do you write up a petition you and the other passengers can sign requesting that the driver cut speed by 20 percent by the year 2025? No, you scratch and claw and kick and bite and do everything you can to get the murderous suicidal asshole's foot off the gas, and press down with everything you've got on the brakes.

❁ ❁ ❁

Let's try this again. This time you're piloting a plane at 30,000 feet, and you smell smoke. A lot of smoke. What do you do? Tell your co-pilot goodbye? Pretend there's no problem? Say there's no evidence that big fires on planes are inherently destructive? No. You try to put out the fire, and you take the plane off autopilot and try to get it on the ground as quickly as possible.

The metaphors should be obvious.

I'm going to extend this metaphor with a story. I sometimes think about the pilots of Alaska Airlines Flight 261 that crashed off the coast of southern California in 2000. While they were in the air, part of the flight control system controlling the pitch of the plane failed, causing the plane to drop nose downward and fall from 31,000 feet to 24,000 feet in about eighty seconds. By pulling hard—130 to 140 pounds of force—on the controls, the pilots were able to stabilize around that latter altitude. They had already discussed making an emergency landing at Los Angeles International Airport. The pilot then wanted to try to coax the jet down to 10,000 feet before attempting a landing. But, and here's the part of the story that always makes me cry, because Los Angeles is so densely populated, the pilot requested permission to try to lose that altitude over the ocean, so a potential crash would kill as few people as possible. The crew got the plane down to 18,000 feet before the crucial screw in the flight control system gave way, and the plane flipped on its back and dove for the ocean. The pilots tried their best for eighty-one seconds, but they hit the ocean at over 150 miles per hour. Everyone on board died.

I'm not too proud to mix metaphors. We're heading toward a brick wall. We need to slam on the brakes as hard as we can. Or if someone else is controlling the speed—if there's a madman behind the wheel—we need to figure out a way to force the brakes ourselves. But that's where the car metaphor fails, because we're not only taking out the passengers, we're taking out everyone else in the vicinity of the crash. So, moving to the second metaphor, if we can't stop or slow the crash, I wish we would at least have the grace and courage of the pilots of Flight 261, and make sure to save as many lives as possible on the way down.

❋ ❋ ❋

But things are far worse than just the authoritarian technics running this culture.

It should also be clear by now that members of this culture, for the most part, cannot even conceptualize living without the benefits they gain from these authoritarian technics, and they have what amounts to no real concern for the victims of the technics, the communities destroyed so they can have their luxuries without which life would be evidently unimaginable. I don't think most people in this culture particularly care if the oceans die, except insofar as it affects their participation in these authoritarian technics (e.g., what does it mean for the economy and my role in it, and most especially, where will I get my fucking sushi?).

But things are still worse. Even the staunchest supporters of this way of life acknowledge (usually without realizing they're doing so) this culture has based itself on overshoot and conquest. We could all become the purest of green pacifists, and the system itself still functionally requires overshoot and conquest. And this basis in overshoot and conquest—along with its associated "virtues" of "growth" and "development of natural resources" and technological escalation—far from being attributes we are collectively even remotely considering abandoning, are instead seen as positive goods. We're ~~ruining~~ running the whole Earth, remember?

But it's still worse even than this, because our human supremacism has long since moved from being an assumption or an attitude or even an unquestioned belief to being our very identity.

This is bad news, indeed.

※ ※ ※

Mumford wrote, "Through mechanization, automation, cybernetic direction, this authoritarian technics has at last successfully overcome its most serious weakness: its original dependence upon resistant, sometimes actively disobedient servomechanisms, still human enough to harbor purposes that do not always coincide with those of the system."[121]

※ ※ ※

Once again, I think it's even worse than this. While mechanization, automation, and cybernetic direction have reduced the dependence of authoritarian technics on resistant humans—far more so now than when Mumford wrote this sixty years ago—I believe that this culture's consumption and destruction of cultural and ecological diversity, and people's declining ability to perceive or imagine a reality outside of this culture, and people's increasing identification with the technics, has greatly reduced our willingness to resist, and our capacity to even conceptualize meaningfully resisting.

❀ ❀ ❀

Mumford wrote, "Like the earliest form of authoritarian technics, this new technology is marvelously dynamic and productive: its power in every form tends to increase without limits, in quantities that defy assimilation and defeat control, whether we are thinking of the output of scientific knowledge or of industrial assembly lines. To maximize energy, speed, or automation, without reference to the complex conditions that sustain organic life, have become ends in themselves. As with the earliest forms of authoritarian technics, the weight of effort, if one is to judge by national budgets, is toward absolute instruments of destruction, designed for absolutely irrational purposes whose chief by-product would be the mutilation or extermination of the human race. Even Ashurbanipal and Genghis Khan performed their gory operations under normal human limits."[122]

❀ ❀ ❀

I have two responses to this. The first is to point out, regarding the weight of effort in this culture being aimed toward destruction, that members of this culture designed and built atomic bombs before they designed and built automatic washing machines. So, judging by where they spend the most money, and where they spend it first . . .

The second has to do with his comment about how even Ashurbanipal and Genghis Khan performed their slaughters under normal human limits. Recall that the writers and editors for *The Atlantic* called gunpowder one of humanity's greatest achievements because it "outsourced killing to a machine."

Are you starting to see how this all fits together? Do you begin to see how this culture is at war with the world itself, with life itself?

❀ ❀ ❀

Actually, I have a third response. As I write this, it is midterm election season in the United States, which means sure as shootin' it is prime time for the United States to bomb and invade more countries. Is there a more reliable way for US politicians to increase their popularity than by killing people, especially when they are able to outsource this killing to machines?

Who's in charge?

❀ ❀ ❀

Mumford continues, "The center of authority in this new system is no longer a visible personality, an all-powerful king: even in totalitarian dictatorships the center now lies in the system itself, invisible but omnipresent: all its human components, even the technical and managerial elite, even the sacred priesthood of science, who alone have access to the secret knowledge by means of which total control is now swiftly being effected, are themselves trapped by the very perfection of the organization they have invented. Like the pharaohs of the Pyramid Age, these servants of the system identify its goods with their own kind of well-being: as with the divine king, their praise of the system is an act of self-worship; and again like the king, they are in the grip of an irrational compulsion to extend their means of control and expand the scope of their authority. In this new systems-centered collective, this pentagon of power, there is no visible presence who issues commands: unlike Job's God, the new deities cannot be confronted, still less defied. Under the pretext of saving labor, the ultimate end of this technics is to displace life, or rather, to transfer the attributes of life to the machine and the mechanical collective, allowing only so much of the organism to remain as may be controlled and manipulated."[123]

CHAPTER SEVENTEEN

SUPREMACISM

I think a lot of self-importance is a product of fear. And fear, living in sort of an un-self-examined fear-based life, tends to lead to narcissism and self-importance.

MOBY

With respect to the idea of possession, I think that with this kind of person [serial sex killers], control and mastery is what we see here. . . . In other words . . . people who take their victims in one form or another out of a desire to possess and would torture, humiliate, and terrorize them elaborately—something that would give them a more powerful impression that they were in control.

SERIAL SEX KILLER TED BUNDY

Every supremacism is unique. Human supremacism is not the same as male supremacism, which is not the same as white supremacism.

However, they do share characteristics. One of these characteristics is their requirement that the supremacists create the perception—and a mythology to promote that perception—that members of classes or groups or sub-groups other than their own are inferior.

And one of the best ways to create and then validate the notion that you are superior to these others is by violating or exploiting them. Surely you couldn't violate or exploit them if you weren't superior, right?

There's a sense, then, in which supremacism is always other-oriented. Not in the sense of caring for or about others, but instead in the sense that supremacism is always comparative, because we can't be superior by ourselves. In this sense, supremacism is about classifying those who are immediately identifiable as different—such as those of different sex, different color of skin, different economic or social class, different species—as inferior to us, making up rationales (from the religious to the scientific to the economic to whatever the hell we can conjure up) for our perceived superiority, and then victimizing these "inferior" others. This victimization then validates our perception of these others as inferior, thereby reinforcing our own sense of superiority to them. Violation becomes not merely an action, but an identity: who we are, and how we and society define who we are. Each new violation then reaffirms our superiority, as through these repeated acts of violation we come to perceive each new violation as reinforcement not only of our superiority over this other we violated, but as simply the way things are.

So without this identification of others as inferior, without this violation, we are not. We are a void. And so we must fill this void, fill it with validations of our superiority, fill it with violations. Thus the violation of every boundary set up by every Indigenous culture. Thus the extinctions. Thus the insane belief in an economic system based on infinite growth despite the fact that we live on a finite planet. Thus the refusal to accept any limits on technological escalation or on scientific "knowledge." Thus the sending of probes to penetrate the deepest folds of the ocean floor. Thus the bombing of the moon. Thus the obsessive repetitions of our claims to superiority. By sufficient repetition of both the violations and the claims of superiority we hope to convince ourselves.

What makes this problem even worse is that because there are always those who have yet to be violated, and because this violation isn't really solving the needs it purports to meet, this drive to violate is insatiable. This culture will continue to violate, until there is nothing left to violate, nothing left.

So what is at stake in this whole discussion is life on this planet. This cult of human supremacism must not merely be left, and must not merely be exposed. Those of us who are not supremacists must stop the supremacists from violating their way to the end of all that is alive.

※ ※ ※

Mumford wrote, "Why has our age surrendered so easily to the control-lers, the manipulators, the conditioners of an authoritarian technics? The answer to this question is both paradoxical and ironic. Present day technics differs from that of the overtly brutal, half-baked authoritarian systems of the past in one highly favorable particular: it has accepted the basic principle of democracy, that every member of society should have a share in its goods. By progressively fulfilling this part of the democratic promise, our system has achieved a hold over the whole community that threatens to wipe out every other vestige of democracy.

"The bargain we are being asked to ratify takes the form of a mag-nificent bribe. Under the democratic-authoritarian social contract, each member of the community [by which he means the global elite] may claim every material advantage, every intellectual and emotional stimulus he may desire, in quantities hardly available hitherto even for a restricted minority: food, housing, swift transportation, instantaneous communi-cation, medical care, entertainment, education. But on one condition: that one must not merely ask for nothing that the system does not pro-vide, but likewise agree to take everything offered, duly processed and fabricated, homogenized and equalized, in the precise quantities that the system, rather than the person, requires. Once one opts for the system no further choice remains. In a word, if one surrenders one's life at the source, authoritarian technics will give back as much of it as can be mechanically graded, quantitatively multiplied, collectively manipulated and magnified."[124]

* * *

I am begging you to reread that last paragraph. Read it again. Put down this book. Wait a couple of days. Read it again. Look at this culture.

* * *

Mumford also said, "I would die happy if I knew that on my tomb-stone could be written these words, 'This man was an absolute fool. None of the disastrous things that he reluctantly predicted ever came to pass!'"[125]

Unfortunately, Mumford was no fool, and even more unfortunately, things are much worse than I'm sure even he ever imagined.

❊ ❊ ❊

He wrote, and this is the last I shall quote him at length in this book, "'Is this not a fair bargain?' those who speak for the system will ask. 'Are not the goods authoritarian technics promises real goods? Is this not the horn of plenty that mankind has long dreamed of, and that every ruling class has tried to secure, at whatever cost of brutality and injustice, for itself?' I would not belittle, still less deny, the many admirable products this technology has brought forth, products that a self-regulating economy would make good use of. I would only suggest that it is time to reckon up the human [and I would add nonhuman] disadvantages and costs, to say nothing of the dangers, of our unqualified acceptance of the system itself. Even the immediate price is heavy; for the system is so far from being under effective human direction that it may poison us wholesale to provide us with food or exterminate us to provide national security, before we can enjoy its promised goods. Is it really humanly profitable to give up the possibility of living a few years at Walden Pond, so to say, for the privilege of spending a lifetime in Walden Two? Once our authoritarian technics consolidates its powers, with the aid of its new forms of mass control, its panoply of tranquillizers and sedatives and aphrodisiacs, could democracy in any form survive? That question is absurd: life itself will not survive, except what is funneled through the mechanical collective. The spread of a sterilized scientific intelligence over the planet would not, as Teilhard de Chardin so innocently imagined, be the happy consummation of divine purpose: it would rather ensure the final arrest of any further human development."[126]

Far more to the point, it would kill the planet.

❊ ❊ ❊

And even more to the point, it already is killing the planet.

❊ ❊ ❊

Today I read an article in *The New York Times* entitled, "In Alaska, a Battle to Keep Trees, or an Industry, Standing." The article details how in Alaska's Tongass National Forest, the United States Forest Service is planning on putting out its biggest timber sale—read: giveaway—in more than ten years: 9.7 square miles of old growth forest, home to many endangered species, such as the Alexander Archipelago wolf (population less than 1300, and falling). The Forest Service has plans to "sell"—read: give away—another eleven square miles of ancient forest in the near future.

Why?

Well, we know the answer to that: this culture hates the natural world, and is doing everything it can to destroy life on this planet.

Let's ask again: why?

Because this culture is enslaved to authoritarian technics. Even the title of this article gives that away, pretending that in decision-making processes an industry should, for some insane reason, be given equivalent weight to the life of a biome.

Of course representatives from the Forest Service would not answer this honestly.

So now let's ask it in a way they *can* possibly answer honestly: what excuses are they using in order to destroy these forests?

"The Forest Service argues that it must keep southeast Alaska's loggers and sawmills in business until a replacement source of timber is ready: second-growth forests, now maturing on lands where virgin forests were clear-cut." In other words, the Forest Service and timber industry are doing what the Forest Service and timber industry do, which is to deforest a region, and then when the region is deforested, use this prior deforestation as one of their many excuses for even more deforestation. The article cites the Tongass Forest supervisor, a more-or-less standard tool of the timber industry named Forrest Cole, as saying, "The industry here is quite small today, and it is kind of on the edge of existing or not. And if we lose it, this whole idea of a transition to a new young-growth industry will probably fail immediately."

Never mind the endangered species and the natural communities who are "on the edge of existing or not." They are never of primary importance to these people.

I want to point out a few things. The Forest Service and the timber industry have already deforested more than 700 square miles in the Tongass National Forest. These timber "sales" have been, as I alluded to, giveaways, with the Forest Service selling old growth trees for less than the price of a hamburger. No, I'm not making that up. These timber giveaways have not only devastated the region, but have cost US taxpayers over a billion dollars. Nearly all of these trees, and nearly all of these subsidies, have gone to two—count 'em, two—huge corporations. These two corporations conspired to drive countless small family handloggers out of business. These corporations have also, no surprise, been consistent polluters. Most of the trees have either been pulped (so, yes, members of this culture have been wiping their asses on toilet paper or looking up pizza delivery places in Yellow Pages made with old growth trees from this region) or sent unmilled to foreign (mainly Japanese) markets. In order to destroy these forests, the Forest Service has punched in more than 4,500 miles of roads. Of course, US taxpayers paid for those roads. And of course, the nonhumans who live in the area have paid with their lives for those roads. But none of that matters, really, to human supremacists. None of it ever matters to them. What matters to them is slavishly serving the technics, in this case getting out the cut.

I want to point out something else: 94 percent of the old growth on Prince of Wales Island, where this timber "sale" is planned, has already been cut. This "sale" would cut into those remaining margins.

And I want to point out something else: because of prior deforestation, the timber industry in that part of Alaska at this point only employs about 200 people.

Even excluding harm to the real world, the current subsidy amounts to about $130,000 per worker.

We'd all be better off if the US government just handed them $100,000 each and told them to stay home.

Or how's this for an idea? Pay them to help the forest to heal. Or an even better idea: since they've already shown themselves willing to destroy the forest to make money—shown themselves willing to destroy forests at all—pay someone else, perhaps someone who loves forests, $100,000/year to actually help the forest to heal. I know plenty of people who could use the money, and who are hard workers. And most importantly, they actually love forests.

But that won't work, since it would violate the first principle of both free market capitalism and of authoritarian technics, which is that destructive activities must have priority in receiving handouts.

Environmentalists are suing the Forest Service to stop the "sale." One of their primary arguments is that the "sale" could drive the Alexander Archipelago wolf closer to extinction.

The New York Times article gives Forest Service Ranger and timber industry shill—but I guess that's nearly always redundant, isn't it?—Rachelle Huddleston-Lorton the last words on why the sales must go forward: "Without the mills, there's no timber industry, and without the Forest Service's second-growth sales, there are no mills. We've got to keep the mills alive."[127]

❅ ❅ ❅

Yes, she actually said that. "Keep the mills alive." "Alive." Of course she did. Not wolves. Not forests. Not salmon. Not living beings. Not biomes. Mills.

Are we all starting to see how authoritarian technics control society?

❅ ❅ ❅

Recall Lewis Mumford's words: "As with the earliest forms of authoritarian technics, the weight of effort, if one is to judge by national budgets, is toward absolute instruments of destruction, designed for absolutely irrational purposes whose chief by-product would be the mutilation or extermination of ~~the human race~~" life on earth.

❅ ❅ ❅

Also today, I read an article on *Truthout* entitled, "Winged Warnings: Built for Survival, Birds in Trouble from Pole to Pole." Some of the article contains important information about the collapse of the biosphere, but human supremacism can't help but raise its narcissistic head, as the very first quote in the article reads, "If birds are having issues, you have to think about whether humans are going to have issues too." Later, a section of the article called "Proxies for People" states that "studies on how endocrine-disrupting chemicals affect birds is a main plank of future research that may also have implications for human health."[128]

Or gosh, maybe if we really cared about human health—never mind the rest of the world—we'd disallow the manufacture of endocrine-disrupting chemicals. But that can never happen, because it would go against the authority and primacy of the technics over our lives.

Then I read another article today about how wildlife populations worldwide have collapsed by 52 percent in the past forty years. Of course this horrifies me. It horrifies me even more that this collapse has come on top of other collapses on top of other collapses on top of other collapses, which means current populations are the merest fractions of what they once were. The world used to be what is now unimaginably fecund. We are witnessing (and as a culture, causing; and as an environmental resistance, doing nowhere near enough to stop) pretty much the final despoliation of this once-vibrant planet. This horrifies me. It all evidently horrifies the journalist, but seemingly for a different reason. Once again, the *very first* quote used in the article is, "'It's the loss of the common species that will impact on people. Not so much the rarer creatures, because by the very nature of their rarity we're not reliant on them in such an obvious way,' said Dr. Nick Isaac, a macroecologist at the NERC Centre for Ecology & Hydrology in Oxfordshire. He says that recent work he and colleagues have been doing suggests that Britain's insects and other invertebrates are declining just as fast as vertebrates, with 'serious consequences for humanity.'"[129]

So here's my question: does *everything* have to come down to how it affects humans? Can we not talk for even a few hundred words about the extirpation of huge swaths of life on this planet without making it all about us? "I'm so sorry you're dying because I'm overworking and starving you. When you're dead, who's going to cook me dinner?"

Why can't these people just want to save these nonhumans from this horrible culture because it is the right thing to do? Why can't we save them out of love? Why can't we save them because they are important to the earth? Why can't we just save them? Full stop. End of sentence. End of paragraph.

❀ ❀ ❀

Why? Because this culture is as narcissistic as it is possible to be. Nothing matters to members of this culture except as it affects them.

❀ ❀ ❀

Oh, yeah, I forgot. We're the ones who developed cooperation . . .

❀ ❀ ❀

I've never really liked the cliché, "You can't solve a problem using the same mindset that created it." But the cliché does get thrown around a lot. Why not here? For once, it's actually relevant.

❀ ❀ ❀

A few days ago someone sent me a note asking for help with a presentation he's doing soon. He wanted a list of ten facts showing pending (or I would say current) ecological collapse. I made him a list—ocean fish reduced 90 percent by weight in the last 140 years; native forests reduced by 98 percent; and so on—and gave it to him. Anyone could easily assemble a list like this. The information is out there. The most difficult part of assembling the list is dealing with your broken heart.

This afternoon he responded with a note that raises everything I'm talking about: "What I'm looking to do is build an argument for devastating ramifications to *us*. A counterargument could be: who cares if fish are down, native forests are two percent, if the plastic goes into the ocean—none of that affects me. There's still food in the grocery store, Facebook, video games, and oh, the Kardashian wedding was so cool! Or look how sunny and nice it is. I went for a nice nature walk today and yesterday, the beach was beautiful. I think a lot of people cannot make the leap from a statistic of how something they cannot see (or verify) [although, of course, to see it, all you have to do is pay attention] to 'we're fucked.' I want to gather some documents to prove a point: that our future is in jeopardy."

I have to admit I've lost all patience with this culture's narcissism. Here is what I want to say: "Honestly, when we're talking about fish in the oceans going extinct we're talking about a larger problem than this simply affecting humans. And since humans are causing the problem I'm not so interested in protecting humans from the rebounding effects of their own actions; it's a bit late in the day to be concerned about the effects of these atrocities on the perpetrators. Human supremacism is the biggest problem facing the planet today, and I don't want to reinforce this culture's narcissism by making it all once again all about precious little us. And if people are too stupid to figure out the relationship between the oceans being dead and their own future, they don't deserve to continue on this still-beautiful planet."

❁ ❁ ❁

A couple of months ago I received an email with the subject header: "Who cares if there are no salmon?" In it, the person said that she herself is concerned about salmon, as are members of her family, but her son also said, "If you told almost everyone that we could save wild salmon from extinction but they would have to sacrifice the things that dams provide, they would say 'who cares about the salmon? So we have no salmon. We need electricity—we don't need salmon.'"

Her son is right about people's response, and about our enslavement to authoritarian technics.

That's why I hate this culture.

She wanted to know what I would say to people who say that. Well, first, I don't talk to them. You can't often argue someone out of bigotry or narcissism or addiction. But if I were going to respond to them, I would say, "Forests care if salmon die. Salmon care if salmon die. Lampreys care if salmon die. Redwoods care if salmon die. Lots of creatures depend on salmon. Salmon help the entire region where they live. On the other hand, who cares if you die? What use are you to the real world? At least salmon help forests, which is more than can be said for most humans. Is the world a better place because you are alive? Not, is this culture better off? Not, have you put in a really nice garden? Not, have you raised children? I'm talking about wild nature. The real, physical world. The real, physical world is better because salmon are in it. The same can't be said of people who prefer industrial electricity to salmon. And for the record, we don't need industrial electricity. We need clean air, water, and food. And habitat. Not industrial electricity."

CHAPTER EIGHTEEN

THE SOCIOPOCENE

I think scientific arrogance really does give a great degree of distrust. I think people begin to think that scientists like to believe that they can run the universe.

ROBERT WINSTON

To be men, we must be in control. That is the first and the last ethical word.

THE NEW ENGLAND JOURNAL OF MEDICINE

The point of science—and this may or may not be true of individual scientists—is to make the world subject to human domination. If they can abstract, and then they can predict on the basis of that abstraction, then they can try, at both the human and natural levels, to use that prediction in order to exert control.

STANLEY ARONOWITZ

Members of this culture are so narcissistic that they're now calling this era the *Anthropocene*: the Age of Man.[130] The term was devised by someone who meant it pejoratively, that humans have become so destructive of the planet that they could be considered a geologic force. But it didn't take long for human supremacists to turn the term into the sort of self-congratulatory rationalization for further destruction to which we have become so accustomed, as in Emma Marris proclaiming we run the earth, as in Charles Mann declaring that "Anything goes."

I find the term really harmful, for a number of reasons, primarily that the term *Anthropocene* not only doesn't help us stop this culture from killing the planet, it contributes directly to the problems it purports to address.

It's also grossly misleading. Humans aren't the ones "transforming"—read: killing—the planet. Civilized humans are. There's a difference. It's the difference between old growth forests and New York City; the difference between flocks of passenger pigeons so large they darkened the sky for days at a time, and skies full of airplanes; the difference between sixty million bison and pesticide- and herbicide-laden fields of genetically modified corn. It's the difference between rivers full of salmon and rivers killed by hydroelectric dams. It's the difference between cultures whose members recognize themselves as one among many, and members of this culture who convert everything to their own use.

Among other problems, the word *Anthropocene* is an attempt to naturalize the murder of the planet by pretending that the problem is "man" and not this particular culture (and other civilizations).

It also manifests the supreme narcissism that has characterized this culture from the beginning. Of course members of this culture would present their own perspective and their behavior as representing "man" as a whole; other cultures have never really existed anyway, except as lesser breeds in the way of members of the One True Way getting access to resources. And of course this is happening today, as Indigenous peoples are still being driven off their land, and Indigenous languages are being driven extinct at a proportionally faster rate than nonhuman species.

Using the term *Anthropocene* feeds into that narcissism. Gilgamesh destroyed a forest and made a name for himself; this culture destroys a planet and names a geologic age after itself. What a surprise.

We've had six thousand years to recognize this pattern of genocide and ecocide committed by members of this culture because of this culture's narcissism, sociopathy, and entitlement, and the behavior is simply getting worse. And members of this culture have had six thousand years to recognize that the cultures they're conquering have often been sustainable. And still they come up with this name that attempts to include all humanity in their own despicable behavior. What a surprise.

The narcissism extends beyond disbelieving that other cultures exist. It extends to believing no one else on the planet fully subjectively exists.

Of course members of this culture, who have previously named themselves with no shred of irony or shame (or humility) *homo sapiens sapiens*, would, as they murder the planet, declare this the Age of Man.

No one else matters, nothing else truly exists. It's like the Catholics say, *Nulla salus extra ecclesium*, which means "Outside the Church there is no salvation." Likewise, this culture believes, "Outside Science there is no knowledge," "Outside Technology there is no comfort," "Outside Capitalism there are no economic transactions," "Outside Industrial Civilization there is no humanity," "Outside civilized humans there is no intelligence," and indeed, "Outside humans there is no function, no purpose, no meaning," and "Outside humans there is no meaningful existence." That's how people can with a straight face insist there's no evidence that agriculture is inherently destructive: there was nothing there to destroy in the first place. That's how Emma Marris can say the world is a giant garden: there was no order until we ordered it. That's how human supremacists can claim to be able to manage a landscape. The forest or wetland or prairie was, to use their words, "decadent" or "inefficient," or "going to waste" before civilized humans—the bearers of all meaning and function—arrived on the scene; it's quite extraordinary that oceans and forests and lakes and rivers survived for millions of years without our assistance. That's how human supremacists can call agriculture a "beneficial use" and act as though water in a river is wasted.

We are the only ones who exist.

According to human supremacists, the world is a giant *tabula rasa* onto which we inscribe our ~~greatness~~ loneliness.

I also have problems naming an age after a mass murderer. Should we call the twentieth century the Age of Hitler? The Age of Stalin? Why don't we extend this to other types of killers and call the 1980s the Age of Ted Bundy?

I'll relent on the question of naming this era after a mass murderer, and I'll even relent on calling this era after this culture. But *Anthropocene* gives no hint of the horrors this culture is inflicting. "The Age of Man? Oh, that's nice. We're number one, right?" Instead, the name must be horrific, it must be accurate, and it must produce shock and shame and outrage commensurate with this culture's atrocities: it *is* killing the planet, after all. It must call us to differentiate ourselves from this culture, to show that this label and this behavior do not belong to us. It must call us to show that we do not deserve it. It must call us to say and mean, "Not one more Indigenous culture driven from their land, and not one more species driven extinct!" It must call us to fury and revulsion. It must call us to use our lives and if necessary our deaths to stop this insane culture from killing all we hold dear, from killing this planet that is the source of all life, including our own.

If we're going to name this age after this culture, we must be honest, and call it The Age of the Sociopath. The *Sociopocene*.

And then we need to end this fucking age, as quickly as possible, using whatever means are necessary.

❊ ❊ ❊

Today I read an Op-Ed in *The New York Times* entitled "Building an Ark for the Sociopocene." No, I lied. It was entitled "Building an Ark for the Anthropocene." But can't you imagine how the article might have read were it accurately titled?

The article begins, "We are barreling into the Anthropocene, the sixth mass extinction in the history of the planet. A recent study published in the journal *Science* concluded that the world's species are disappearing as much as 1,000 times faster than the rate at which species naturally go extinct. It's a one-two punch—on top of the ecosystems we've broken, extreme weather from a changing climate causes even more damage. By 2100, researchers say, one-third to one-half of all Earth's species could be wiped out. As a result, efforts to protect species are ramping up as governments, scientists and nonprofit organizations try to build a modern version of Noah's Ark. The new ark certainly won't come in the form of a large boat, or even always a place set aside. Instead it is a patchwork quilt of approaches, including assisted migration, seed banks and new preserves and travel corridors based on where species are likely to migrate as seas rise or food sources die out. The questions are complex. What species do you save? The ones most at risk? Charismatic animals, such as lions or bears or elephants? The ones most likely to survive? The species that hold the most value for us?"[131]

The article goes on to describe some of the efforts, which are of course desperately important, and some of the ways different people and organizations can make these difficult decisions. There's a part of me that is happy that the corporate news is taking time out of its busy schedule to mention the murder of the planet. After all, these 1,200 words could have been used to cover other topics, like someone's folksy reminiscences of gummi bears, or someone else's analysis of how "Ladyfag is the rave of the future," or the extremely important information that the stock market dropped sharply today over fears that the economy isn't growing fast enough.[132] Such is the poverty of our discourse that mere mention of the biggest problem the world has ever faced can be enough to make me, well, *happy* isn't the right word. . . . Perhaps grateful, like a starving dog thrown the tiniest crust of bread.

Not surprisingly, though, my response is mixed. My first problem is that this is precisely where this culture has been headed since its beginnings: it has *always* wanted to play God and decide who lives and who dies. That's a central point of human supremacism. How do we know we're superior? Because we're the ones who are deciding. We're the ones who do *to*, as opposed to everyone else, to whom it is done. We're the subjects. They're the objects. From the beginning members of this culture have wanted to be God. That is, they've wanted to be the God they created in their own image. That is, the God created in the image of how they wanted to be—omnipotent and omniscient—and in the image of how they themselves actually were: jealous, angry, abusive, vengeful, patriarchal. It pleases the supremacists no end to pick up the civilized man's burden and pretend they're being merciful in deciding which of their lessers to exterminate, and which to save. For now.

But there's a much bigger problem than this. Did you notice who is on the chopping block, and what is not? Did you see it?

What is missing is any mention of technics, technologies, luxuries, comforts, elegancies. Sure, we're supposed to choose whether to extirpate or save Bulmer's fruit bats or Sumatran Rhinos, wild yams or hula painted frogs (with the default always being extirpate, *of course*); and we're supposed to make careful delineations of *how* we choose who is exterminated, and who lives (at least until tomorrow, when we all know there'll be another round of exterminations, complete with another round of wringing our hands over how difficult these decisions are, and another round of heartbreak; and then another round, and another, until there is nothing and no one left); but just as the Japanese energy minister said that no one "could imagine life without electricity," so, too, entirely disallowed is any discussion of what technologies should be kept and what should be caused to go extinct. There's no discussion of extirpating iPads, iPhones, computer technologies, retractable stadium roofs, insecticides, GMOs, the Internet (hell, Internet pornography), off-road vehicles, nuclear weapons, predator drones, industrial agriculture, industrial electricity, industrial production, the benefits of imperialism (human, American, or otherwise).

Not one of them is mentioned. Never. Not once.

Why? Because we are God and God never relinquishes power. We are omniscient and omnipotent, and we are the top of the pyramid. We are the champions, and we can and will do whatever the fuck we want.

None of these are mentioned because none of the benefits of our dismantling of the planet can be seriously questioned.

Anti-imperialist discourse provides a great example of this lack of serious questioning. Of course anti-imperialists rail against imperialism—that's what anti-imperialists *do*—but so many of them don't seem to understand that you can't have the benefits of imperialism without having the imperialism itself. So they will argue against imperialism at the same time that they argue in favor of, for example, high speed rail or groovy solar panels. But you can't have high speed rail and groovy solar panels without mining and transportation and energy infrastructures, and you can't have those infrastructures without the military and police to control them. And in terms of the planet, you can't have any of those infrastructures without the harm those infrastructures and their related activities cause. And since almost none of the anti-imperialists will question those basic infrastructures, that means most of them aren't, in all truth, questioning the imperialism.

Here's how it works regarding this ark for the Sociopocene: we gain the benefits, and now we're pretending that we face this terrible dilemma as to which of our victims we're going to save (for now). But that's not really a dilemma. Let's pretend I'm going to kill either you or your best friend. And no matter whom I kill, I'm going to take everything you both own and everything you hold dear. I gain and both of you lose, including, for one of you, your life. I choose which one dies. That's not a dilemma for me. To qualify as a dilemma I have to have something at stake. Instead of a dilemma, it's murder and theft.

But from a supremacist perspective, I'm not a murderer and thief. I'm a savior. I saved one of you from certain death (admittedly, at my own hands, but still). And being this savior is more evidence of my superiority. A lesser being might have mindlessly killed you both. Gosh, aren't I great? And since I'm so smart, maybe I can come up with all sorts of criteria by which today I'll make my decision as to which of you I'll kill. Then tomorrow, I'll make another decision based on these or whatever other criteria I want as to whether to kill the survivor from today or your second-best friend. And the day after, I'll make this decision again with someone else you love.

I find it deeply troubling that at least some members of this culture can feel even remotely good about themselves for choosing who lives and who dies, if they don't also work toward stopping the actual cause of the murders. It's analogous to a guard at a Nazi death camp feeling like a hero for giving Sophie the choice as to which of her children he won't murder (tonight).

Once again, the murder of the planet is not some tragedy ordained by fate because we're too damn smart. It is the result of a series of extremely bad social choices.

We could choose differently. But we don't. And we won't. Not so long as the same unquestioned beliefs run the culture.

Don't get me wrong. Anyone who is working to protect wild places or wild beings from this omnicidal culture is in that sense a hero. We need to use every tool possible to save whomever and wherever we can from this culture. But it's ridiculous and all-too-expected that while there's always plenty of money to destroy the Tongass and every other forest, and there's always plenty of money for various weapons of mass destruction (such as cluster bombs or dams or corporations), somehow, when it comes to saving wild places and wild beings, we have to pinch pennies and "make difficult decisions."

Also, I need to say that the whole ark metaphor doesn't work. In the original story, God saved two of every species (as He, like the humans who created Him, destroyed the planet). Here, modern humans are going where even God didn't tread, and explicitly not saving every species, but instead deciding which species to save, and which species to kill off. This is, of course, both pleasing and flattering to human supremacists: they're making decisions on questions even God punted.

How cool is that?

There's an even bigger problem than all of these, though, which is that this culture is systematically and functionally killing the planet. All the wonderful and necessary work of every activist who is fighting as hard as she or he can to protect this or that wild place won't mean a fucking thing so long as this culture stands. And all this fine work that goes into creating decision-trees as to whom we deem worthy of saving and whom we will drive extinct is meaningless when we completely fail to address the cause of the murders in the first place.

Until civilization collapses, the murder of the planet won't stop.

Picture this. A gang of sadistic, vicious, insane, entitled, sociopathic murderers has taken over your home, and is holding everyone you love captive. They are systematically pulling your loved ones from the room and torturing them to death. What do you do? Do you make decision-trees to help you make "difficult decisions" as to which of your loved ones you'll hand over next? Maybe you do. But I have to tell you I'd be more focused on stopping the murderous motherfuckers in their tracks, stopping the murders at their source.

From the perspective of human supremacists, though, it is easier, more pleasing, and certainly reinforces one's own identity as superior, to "reluctantly" make "difficult decisions" as to who will be driven extinct. So long as we never, ever, ever question the supremacism and the culture that is driving them extinct. And so long as we never forget to go along with Mumford's "magnificent bribe."

We know on which side our bread is buttered.

※ ※ ※

Let's drop the rhetoric. The op-ed broke my heart, not only because the murder of the planet breaks my heart; and not only because the op-ed discussed which creatures to let go drive extinct without talking about which technics to let go get rid of; and not only because, of course, they mentioned which species are most useful to us, but entirely absent among their criteria for saving species was that of which beings best serve life on earth (and, of course, missing entirely was any discussion of what technics serve life and what harm life); but even more so because it completely ignored what is in many ways the only thing that matters: stopping the primary damage. The truth is that these other beings wouldn't need to be saved if civilization wasn't killing them. The truth is that they can't be saved so long as civilization is killing the planet. And the truth is that in this culture there are certain topics which must never be discussed, certain self-perceptions and perceived entitlements which are never negotiable. We would rather kiss ourselves and the entire planet good-bye than to look honestly at what we have done, what we are doing, and what we will, so long as we have this supremacist mindset, continue to do.

✿ ✿ ✿

Another big problem with the idea of an ark for the Sociopocene is that it's based on and promotes this culture's harmful and inaccurate view of the natural world, that you can take a creature out of its habitat and still have the complete creature, that a prairie dog is just a bundle of DNA in a fur and skin sack, and not part of the larger body of the prairie.

This culture seems to believe—completely anthropomorphically—that the world is like a machine or a chair. Some human artifact. Something where the whole is no more than the sum of the parts. You can take apart a chair and swap out some parts, then put the chair back together, and you still have a chair (except that this culture would steal a bunch of screws, two legs, and the seat, then wonder why they can't sit in it; but don't worry, it's just been "reorganized"). But that's not how life works, whether we're talking about a human body or the body of a river or a prairie. The whole is more than the sum of the parts. And if you don't think so, have a surgeon take you all apart and put you back together. Call me when they're done. I'll have my Ouija board set on vibrate.

You can't remove a wolverine from its habitat and still have a wolverine. You have something that looks and smells like a wolverine. But the wolverine is also the scents it picks up on the breeze and the soil under its feet. Without the weather patterns and everything else about where it lives, it would not have become the being it is.

Yes, the Franklin's bumblebee must be saved, as must the Hunter's hartebeest and the Chinese bahaba and the Galapagos damselfish.

But they don't need arks. What they need is a living planet. What really need to be protected are the larger bodies who are their homes, the oceans, the forests, the rivers, the lakes, the entire larger communities.

❀ ❀ ❀

I'm not saying we should never intervene. Obviously every being intervenes at all times: salmon affect forests, trees affect salmon, prairie dogs bring rain (and no, I'm not being hyperbolic; they do). What I'm saying is that we learn to play our proper role, and that we intervene with humility. Charles Mann was wrong when he said that "Anything goes." We need to act in ways that improve the health of the land on its own terms.

I live in far northern California. The local National Forest District is one of the most well-managed I've ever encountered. Why? Because there is no timber sale program, so here the Forest Service has as one of its primary goals not the theft of trees to serve the timber industry, but rather the righting of old wrongs, such that the District removes a lot of old mining and timber roads, and works to help the local rivers and forests the best it can.

Likewise I have a friend who teaches at a state university in New York. The university "owns" a forest. Unfortunately, because wolves and mountain lions have been eliminated from the region, the forest is being overrun with white-tailed deer. Neither young trees nor underbrush survive, which also means amphibians and rodents and ground-dwelling birds don't survive. To save the forest, the university needs to either kill some deer, or better, reintroduce mountain lions or wolves.

Lots of human supremacists say that because American Indians affect the land where they live, that somehow this means that "anything goes." But that's simply an excuse for abhorrent behavior. The truth is that Indigenous peoples traditionally lived in place, and made decisions with the understanding that they were going to be living in that place for the next 500 years. If you are planning on living in place for the next 500 years, your community will make far different land-use decisions. You won't destroy the rivers. You won't destroy the forests or grasslands. You won't drive off predators. You won't poison the land, water, and air. You won't harm the natural communities of whom/which you are a part.

It is not okay to manage a natural community for extraction. It is not okay to manage a natural community for a sense of self-aggrandizement. It is acceptable to try, with humility and with the understanding that you are going to be living in place for the next 500 years, to right the wrongs of the human supremacists who surround you.

And, of course, it is acceptable to stop these human supremacists from causing more harm.

❖ ❖ ❖

I recently did a benefit for the Buffalo Field Campaign, a great organization trying to protect the last free-ranging herd of genetically pure bison in the United States, in Yellowstone National Park. In order to serve the financial interests of a few cattle ranchers, the Montana Department of Livestock and the United States Park Service shoot hundreds of these bison each year, and otherwise manage them to death. Buffalo Field Campaign is the first, and often last, line of defense against this manage-ment murder. Mike Mease is BFC's co-founder, and has been doing this work for seventeen years. He's a force of nature.

We were chatting after the benefit, and he said something that ties directly into this book. The Park Service had orphaned a bison calf, and removed it from the herd. Mike asked the Ranger, now that the calf had been separated from his family, who was going to teach the calf how to become a bison. The Ranger said, "*I* will."

This is everything I'm talking about in this book in two small words. He knows how to teach this child to become an adult bison. Of course. Good luck teaching him how to deal with other males during rut.

Perhaps the notion of humans attempting to manage the natural world reveals more than anything else the complete insanity of human supremacism, and this supremacism's near-absolute invulnerability to counter-evidence. This culture has critically harmed or destroyed *every single biome* it has managed, and yet the managerial ethos gets stronger every day. Forests: managed to death. Yet still the managers claim to know what is best for forests. Wetlands: managed to death. Yet still the managers claim to know what is best for wetlands. Rivers: managed to death. Yet still the managers claim to know what is best for rivers. Oceans: managed to death. Yet still the managers claim to know what is best for oceans.

Maybe bison know best how to raise a bison child, and forests know best how to raise and maintain forests, and wetlands know best how to build and maintain wetlands, and oceans know best how to keep themselves alive and healthy. They've been doing this for quite a while.

If a doctor killed or injured *every single patient* he saw, would you trust your life to this doctor? If a cop bungled *every single case* she handled, would you want her investigating (or preventing) the death of your loved one? If *every single bridge* built by a certain engineer collapsed, would you want him building bridges over which you and those you love will travel? If a financial advisor gave you bad advice *every single time* she opened her mouth, would you trust her with your financial future?

Yet this is the track record of human supremacists.

It's of course even worse, because "management" in human supremacist terms really means "stealing as much as possible" from the one being "managed." So as well as being completely insufficient to the task, the doctor is stealing organs from his patients, the cop is busy killing relatives of the dead and cutting rings off their fingers, the engineer not only steals building materials to re-sell, but also loots the homes of those who die in the collapses, and the financial advisor's real goal all along was to gain access to your assets.

In each case, the managers do fine. Their victims, not so much.

In our own lives, no one would entrust anything to these thieves and murderers—to these thieves and murderers who, even if they weren't thieves and murderers, would *still* be incapable of performing up to their claimed abilities—yet, time and again, we entrust the source of all life to them.

Part of the reason, of course, is Mumford's magnificent bribe. We have sold whatever native intelligence and integrity and empathy and common sense we have in exchange for our cut of the swag. I hope you enjoy your share of the money we got for the ring that used to be on our big sister's finger. I know I sure did. I bought myself a new computer. And Mom's liver paid this month's electricity bill.

Another reason we're all so stupid about this has to do with our enslavement to authoritarian technics. The technics say we can manage anything we turn our minds to, and who are we to question the all-wise and all-knowing technics? Technics, by the way, that brought me this damn computer, and electricity itself, without which life would be unimaginable, so you keep your mouth shut about any of your so-called "downsides" to this technology. Jesus. Fucking ingrate.

With apologies again to Upton Sinclair, it's hard to make a man understand something when his entitlement depends on him not under-standing it.

Rationalized theft really is a big part of it. Robert Jay Lifton wrote that before you can commit any mass atrocity, you have to have what he called a "claim to virtue," that is you have to convince others and espe-cially yourself that you are not in fact committing an atrocity, but instead performing some virtuous act. So the Nazis weren't committing mass murder and genocide, but rather purifying the Aryan "race." The Amer-icans weren't committing mass murder and genocide, but rather mani-festing their destiny. And are, of course, still doing so. Members of the dominant culture aren't killing the planet, they are "developing natural resources." And it's not mass murder, theft, and ecocide, it is "managing" forests, wetlands, rivers, and so on.

Lifton was not the first to observe this role of self-delusion in the per-petrating of atrocities. I'm sure some humans have been decrying this as long as other humans have been doing it. In the late fourteenth century, for example, Timur, sometimes known as Tamurlane the Great, initiated military campaigns that killed about seventeen million people, or 5 per-cent of the human population at the time. About this he said, "God is my witness that in all my wars I have never been the aggressor, and that my enemies have always been the authors of their own calamity." Historian Edward Gibbon responds, "During this peaceful conversation [when Timur said this] the streets of Aleppo streamed with blood, and reechoed with the cries of mothers and children, with the shrieks of violated vir-gins. The rich plunder that was abandoned to his soldiers might stimulate their avarice; but their cruelty was enforced by the peremptory command of producing an adequate number of heads, which, according to his custom, were curiously piled in columns and pyramids."[133] Gibbon also comments, with his usual dryness, "For every war a motive of safety or revenge, of honor or zeal, of right or convenience, may be readily found in the jurisprudence of conquerors."[134] And that is certainly true today in the war on the natural world (which means, the war on life itself).

It seems as though most of the time we use most of our intelligence not to solve problems, but rather to rationalize our atrocities. Certainly this is one of the primary functions of Western philosophy, science, religion, economics, popular culture, "news," political theory, and so on.

And yet another reason we keep pretending, against all evidence, that we can manage the natural world (and, in fact, that we keep pretending that evidence showing we can't manage the natural world doesn't exist), has to do with that core unquestioned belief of human supremacism: humans have intelligence, and nobody else does. Humans have function and purpose and meaning, and nobody else does. Remember, the human brain is the most complex phenomenon in the universe. So of course humans can manage a forest, and humans can manage the oceans, and can manage rivers and whatever else we want to manage. How hard can it be?

Well, I'll tell you what would be very hard for us, which would be for us to acknowledge that from the very beginning we have been wrong. If we were to acknowledge that our management failed—and I don't mean just one time where we made mistakes we are learning from and next time will do better, but instead the ubiquitous and functional failure of the entire managerial ethos—we would have to acknowledge that perhaps we aren't as omniscient and omnipotent as we believe. We would have to acknowledge that the totality of our human supremacist mindset is based on the lie that we are intelligent and they are not.

Let's look at the evidence. Who has done a better job at managing forests (rivers, oceans) in the past: forests (rivers, oceans), or human supremacists? Obviously forests (rivers, oceans).

So let's presume for a second that human supremacists are right, and that there is no intelligence in nature. There is no purpose. There is no function. I don't believe any of this for a heartbeat, but for the sake of argument let's grant them this. What, then, does this say about human intelligence that humans do a *worse* job of "managing" forests (and everything else) than does unintelligent, purposeless, functionless nature? Forests survive and in fact become over time more fecund and resilient on their own. They don't survive human supremacist "management." What does that say about what we call our own intelligence? We consistently—*every single time*—do worse than mindless nature. What does that make us?

Maybe that's because wild forests know what they want, and wild rivers know what they want, and wild oceans know what they want, and they all know how to get it, so long as they aren't being murdered. And part of what they want is to not be enslaved, to not be made to jump through hoops on command.

Of course, another reason human supremacists keep telling themselves and everyone else they can manage the world as they destroy the world is that they hate wild nature. They hate and want to destroy all they cannot control, in part because through this destruction of what they cannot control they show, in their own minds, that they are superior to those they are destroying: were I not superior I could not destroy you. And they hate and want to destroy all they cannot control, because what they cannot control reminds them that they are not in fact superior to these others.

The point is theft and murder—the point is violation—and the point is to declare oneself superior for doing so.

Nature doesn't exist, and insofar as it does, we will destroy it. This is the point of human supremacist management. This is why we can't acknowledge that human supremacist management always fails: because, in fact, it doesn't: it achieves what it set out to achieve: the destruction of the biosphere.

❀ ❀ ❀

All of this is why and how Emma Marris and others like her can say there is no wild nature anymore, and this is how and why so many members of this culture can say there are no costs to agriculture.

❀ ❀ ❀

Another bison story, another story of the failure of management. A few decades ago an Indian nation in Montana wanted to conduct a traditional bison hunt. They were mandated to consult with federal managers, who came up with a plan: the Indians were to kill the old bulls, those who were past their sexual prime, and as such, useless in terms of passing on genetic material. Everybody wins: the Indians get their food and hides, the bison herd doesn't lose any necessary genetic materials (because the old bulls were too old to have sex ever again, the bison were, from a strict genetic perspective, already dead bulls walking), the federal managers get to kill some wild nature *and* file paperwork showing tangible actions leading to increased appropriations possibilities in the next fiscal year, and the human supremacists get to feel superior. Nonetheless, the Indians said that this is not what their teachings suggested. They insisted that the bulls had a role as elders in the bison community. The managers were unswayed by this non-scientific argument. In a fight between "teachings" and the tools of scientific management (backed by the full power of the state) scientific management nearly always wins, and the world generally loses. The only way the Indians could have their traditional hunt is if they killed the animals the managers told them to. So they did.

That winter the remaining bison starved. Life is way more complex than managers think it is. It is more complex than any of us think it is. It is more complex than we are capable of thinking. Montana winters are cold and the snow can be deep. Bison need to eat. How do they get through the snow to the vegetation beneath? It ends up that the old bulls are the only ones whose necks are strong enough to sweep away the heavy snow. They do this for their whole community.

As usual, the managers make the decisions, and others pay the consequences.

* * *

Who would be arrogant enough to believe they can understand all of the relationships in some natural community? Who would be arrogant enough to believe that they know better than bison who is and is not crucial to the survival of their bison community?

Yet this is what managers do time and again. They presume to know better than rivers whether rivers need salmon, and whether salmon need rivers. They presume to know better than wetlands what and who the wetlands need. Who can predict the effects of the loss of a certain species of beetle, of a certain strain of bacteria, of a certain fern or reptile or small mammal? Recall the relationships between parasites, fish, and the seabirds who eat the infected fish. I have Crohn's disease. It is a disease of civilization. As countries industrialize, there is a dramatic increase in Crohn's. One of the theories is that an absence of intestinal parasites leads to this and many other autoimmune disorders (the parasites, for the most part harmless in themselves—I've intentionally infected myself with them, and my health has improved because of it—temper our immune system). Who could predict all of these relationships?

From a human supremacism perspective, it doesn't really matter that the managers destroy everything they touch, because even when the human supremacists have the evidence that their actions are harmful, they ignore this evidence. This is something we've seen once or twice, or maybe every moment of every day. The human supremacists seem to believe that their willful ignorance of these harmful consequences means that there are none.

Or maybe they just don't give a shit.

CHAPTER NINETEEN

EARTH-HATING MADNESS

Genocide is not just a murderous madness; it is, more deeply, a
politics that promises a utopia beyond politics—one people, one
land, one truth, the end of difference. . . . Genocide is a form of
political utopia.

MICHAEL IGNATIEFF

I just read interviews with members of the family owning SpaceX (Space Exploration Technologies Corporation), Tesla (electric) Motors, and SolarCity, this latter the largest solar power manufacturing company in the United States. The interviews fit well together, like imperialism and genocide, like the industrial economy and the murder of the planet.

The first is entitled, "Q&A With SolarCity's Chief: There Is No Cost to Solar Energy, Only Savings: How long will the world and the U.S. continue to tolerate being able to pollute for free?"

The article presents two interesting exchanges. In the first, the interviewer asks, "What's your feeling about where the clean energy movement is right now?" and the "Chief" answers, "It's absolutely getting bigger. How long will the world and the U.S. continue to tolerate being able to pollute for free? Every fossil fuel company should have to admit, 'We are allowed to pollute for free.' That pollution is putting a tremendous amount of cost on all these other externalities. That pollution should be included in the cost of the product. People are going to realize that."

I completely agree that "fossil fuel" companies should include pollution in the price of their product, and would extend that to every action done by this culture. This culture is based on "polluting for free." That's how people in this culture can say that agriculture is not inherently destructive. That's how people in this culture can pretend you can have infinite growth on a finite planet. That's how people in this culture can always bring every atrocity committed by this culture back to that most important question: How does my committing this atrocity affect me? If you can pollute for free, baby, the only thing that matters is that you yourself don't have to smell it (or die of cancer from it).

Which brings us to the second quote from this interview.

Question: "Is the upfront cost of solar still the big burden, the big hurdle for most US homeowners?"

Answer: "The biggest burden, quite frankly, is still education. People still associate solar as having a cost. But there is no cost. There's just savings. In all of our products . . . there's essentially no cost to the homeowner. They just save money from day one. . . . Although they're buying the system, and the system may cost $30,000, they're getting a loan for $30,000, and then they're paying back the loan based on the production of the solar system."[135]

Ah, so I see how it works. When someone *else* devastates the natural world, *they're* being allowed to "pollute for free" and that pollution should be included in the cost of the product; but when *I* devastate the natural world, we only talk about how there's no cost to homeowners. Suddenly the pollution shouldn't be included in the cost to the consumers. I guess that's how we can say there's no evidence agriculture is inherently destructive. That's how we can talk about sustainable development.

Unfortunately for the real world, there *are* costs associated with solar photovoltaics. Solar panels, no matter how groovy, require mines. In addition to copper and other metals, the panels require rare earths minerals (used also in cell phones, batteries, wind turbines, and a host of other high-tech devices). Nearly all of these rare earths are mined in China. Nearly half of all rare earths in China are mined near the city of Baotou (the name means, sadly, "the place of deer," but I guess it was named a long time ago), and most of *this* is from one open pit mine more than a half mile deep and covering (or rather uncovering, or rather killing) more than eighteen square miles. The costs don't stop there. Rare earths are found in extremely low concentrations, and must be separated from the rest of the ore. The separation processes require the use of sulfates, ammonia, and hydrochloric acid, and produce 2,000 tons of toxic waste for every ton of rare earths. The mines and smelters and factories of Baotou alone produce ten million tons of wastewater per year. This "water" is pumped into tailings ponds, including one that covers almost four square miles and about which *The Guardian* has written, "From the air it looks like a huge lake, fed by many tributaries, but on the ground it turns out to be a murky expanse of water, in which no fish or algae can survive. The shore is coated with a black crust, so thick you can walk on it." *The Guardian* also wrote, "The foul waters of the tailings pond contain all sorts of toxic chemicals, but also radioactive elements such as thorium which, if ingested, cause cancers of the pancreas and lungs, and leukemia. 'Before the factories were built, there were just fields here as far as the eye can see. In the place of this radioactive sludge, there were watermelons, aubergines and tomatoes,' says Li Guirong with a sigh." The soil and water are so polluted that the local residents can no longer grow vegetables there. Many have fled. Many have been forcibly relocated. Many have died, and those who remain are suffering a host of diseases caused by this mining.[136]

I'm glad there are no costs, only benefits.

And did I mention the slave labor? As Max Wilbert states, "A substantial portion of the Chinese workforce, especially for the dirty jobs like this that are likely to result in cancer, lung disease, or asthma, comes from Tibet, where communities are forcibly disbanded by the Chinese military and sent hundreds of miles from their homes and traditions to work in the coal, uranium, and rare earth mines. A full fifth of Tibet's population has been killed since China's occupation began, with a substantial portion of those worked to death in forced labor camps. At this point that's one point two million people and counting."[137]

There are plenty of other consequences (by all means we should never call them costs), but let's mention only two.

One is that the production of solar panels is a leading source of the potent greenhouse gases hexafluoroethane, nitrogen triflouride, and sulfur hexafluoride; with hexaflouroethane being 12,000 times more potent as a greenhouse gas than carbon dioxide and lasting 10,000 years in the atmosphere (and it does not exist in nature, which I guess means humans really are superior since they made this pollutant); nitrogen trifluoride being 17,000 times stronger than CO_2 (with concentrations rising in the atmosphere at more than 10 percent per year); and sulfur hexafluoride being 25,000 times more powerful than CO_2.[138]

The other certainly-not-a-cost I want to mention is discussed in a report by the Silicon Valley Toxics Coalition: "As the solar industry expands, little attention is being paid to the potential environmental and health costs of that rapid expansion. The most widely used solar PV panels have the potential to create a huge new source of electronic waste at the end of their useful lives, which is estimated to be 20 to 25 years. New solar PV technologies are increasing efficiency and lowering costs, but many of these use extremely toxic materials or materials with unknown health and environmental risks (including new nano materials and processes)."[139]

The second interview was with SolarCity's Chairman, and CEO of Tesla Motors and SpaceX, Elon Musk. Musk is a big deal these days. He has won scads of awards, and in 2010 was listed by *Time* as one of the 100 people who are most affecting the world. *Esquire* named him one of the seventy-five most influential people of the twenty-first century (rather prematurely, I'd think, since we're less than 20 percent of the way through). In 2013 *Fortune* named him businessperson of the year. In 2008 the National Wildlife [sic] Federation gave him their National Conservation Achievement award for his work with Tesla Motors and SolarCity.

Keep that National Conservation Achievement Award in mind[140] as you read the beginning of the interview.

It starts, "'Fuck Earth!' Elon Musk said to me, laughing. 'Who cares about Earth?'"

Well, I do—and I think so does everyone on the planet who isn't a human supremacist—but his question was rhetorical.

The interviewer was quick to clarify that Musk was kidding, that in fact Musk cares a great deal about the earth. How does the interviewer let us know this? He *must* love the earth, because, "When he is not here at SpaceX, he is running an electric car company." Ah, I see, when someone else pollutes, they're polluting for free, but when I pollute there is no cost, and when I run a car company that puts the word *electric* in front of it, it means I love the earth, even when I don't.

SpaceX, or Space Exploration Technologies Corporation, is a "space transport services company" that Musk founded with the goal of colonizing Mars. He's not some lone lunatic. He has an entire culture for company. Recall the awards he has won, including one from an organization—National Wildlife Federation—that raises almost 90 million dollars per year supposedly to protect the earth (Fuck Earth, right?).

SpaceX is a privately held corporation that has 4,000 employees and has as its largest customer NASA (i.e., US taxpayers).

The interview states, and as you read this try to think about how unquestioned beliefs are the real authorities of any culture, "'I think there is a strong humanitarian [sic] argument for making life multi-planetary' [by which he means making human life multi-planetary while this culture kills this planet], 'in order to safeguard the existence of humanity in the event that something catastrophic were to happen [like perhaps the murder of the planet by human supremacists?] in which case being poor or having a disease would be irrelevant, because humanity would be extinct. It would be like, "Good news, the problems of poverty and disease have been solved, but the bad news is there aren't any humans left."'

"Musk has been pushing this line—Mars colonisation as extinction insurance—for more than a decade now, but not without pushback. 'It's funny,' he told me. 'Not everyone loves humanity. Either explicitly or implicitly, some people seem to think that humans are a blight on the Earth's surface. They say things like, "Nature is so wonderful; things are always better in the countryside where there are no people [sic] around." They imply that humanity and civilisation [and please note his conflation of humanity and civilization] are less good than their absence. But I'm not in that school,' he said. 'I think we have a duty to maintain the light of consciousness [sic], to make sure it continues into the future.'"

Fuck Earth indeed.

I, on the other hand, think humans don't have a monopoly on the "light of consciousness," and that we, like everyone else on the planet, have a duty to leave the planet in a better condition than that into which we were born.

The article states, and as you read this you might find yourself unaccountably adopting a tone of awe-filled narcissistic reverence, "Unlike light, whose photons permeate the entire cosmos, human-grade consciousness appears to be rare in our Universe. It appears to be something akin to a single candle flame, flickering weakly in a vast and drafty void."

Of course human supremacists believe "human-grade consciousness" is rare, or better, unique. We only see what we want to see. That's how we know agriculture isn't inherently destructive, and that's how we know there are no costs to solar photovoltaics. And that's how we know only humans are intelligent.

The article: "Musk told me he often thinks about the mysterious absence of intelligent life in the observable Universe."

It's not mysterious at all; he doesn't perceive intelligent life in the observable Universe for the same reason, once again, that there is no cost to solar photovoltaics. He defines human intelligence as the only intelligence, then wonders why only humans manifest intelligence.

I often think about the mysterious absence of intelligence among human supremacists.

He continues, "Humans have yet to undertake an exhaustive, or even vigorous, search for extraterrestrial intelligence, of course. But we have gone a great deal further than a casual glance skyward. For more than 50 years, we have trained radio telescopes on nearby stars, hoping to detect an electromagnetic signal, a beacon beamed across the abyss. We have searched for sentry probes in our solar system, and we have examined local stars for evidence of alien engineering. Soon, we will begin looking for synthetic pollutants in the atmospheres of distant planets, and asteroid belts with missing metals, which might suggest mining activity."

And there you have it: a sure sign of intelligence is pollution. And another is mining, which is the second most destructive human activity after agriculture (which means, of course, that it also must not be inherently destructive). How much clearer does it have to be?

I guess the Indigenous humans who didn't produce air pollution and who didn't mine must not have been intelligent. Just as the nonhumans who don't produce air pollution and who don't mine must not be intelligent.

No wonder some of these human supremacists perceive intelligence as destructive. That's how they've defined it.

To be clear: when scientists search for signs of intelligent extraterrestrial life, they are actually searching for signs of authoritarian technics. But authoritarian technics is not the same as intelligence. Worse, because these authoritarian technics are not sustainable—as even Musk accidentally acknowledges when he speaks of searching for pollution—all these scientists are wasting their time looking for civilizations (not "intelligent life") that couldn't possibly have lasted very long. For all that scientists love to talk about "nature's laws," they seem to think these "laws" don't apply to us (and when do supremacists ever think any laws apply to them?); you can't convert a planet to machines and expect to live on it.

He continues, "The failure of these searches is mysterious [no, it isn't, for the reason I just gave], because human intelligence should not be special. [It isn't.] Ever since the age of Copernicus, we have been told that we occupy a uniform Universe, a weblike structure stretching for tens of billions of light years, its every strand studded with starry discs, rich with planets and moons made from the same material as us. If nature obeys [sic] identical laws [sic] everywhere, then surely these vast reaches contain many cauldrons where energy is stirred into water and rock, until the three mix magically into life. And surely some of these places nurture those first fragile cells, until they evolve into intelligent creatures that band together to form civilisations, with the foresight and staying power to build starships."

And there we go again: intelligence is defined as creating civilizations and starships. Left unsaid is the process of murdering the planet that inheres in civilizations. But the point remains: intelligence is here defined in action as theft from landbases for use by the "intelligent." But that's not intelligence. That's just theft. And when that theft harms the planet that keeps you alive, that's really stupid.

Musk says, reinforcing the Great Chain of Being theme, and once again giving away the whole rotten game that is human supremacism, "At our current rate of technological growth, humanity is on a path to be godlike in its capabilities."

That's what we've wanted from the beginning, right?

He also says, "You could bicycle to Alpha Centauri in a few hundred thousand years, and that's nothing on an evolutionary scale. If an advanced civilisation existed at any place in this galaxy, at any point in the past 13.8 billion years, why isn't it everywhere? Even if it moved slowly, it would only need something like .01 per cent of the Universe's lifespan to be everywhere. So why isn't it?"

Because civilizations aren't sustainable. Ernst Mayr's comment about intelligence being a lethal mutation would be true if we changed it to: Authoritarian technics are lethal social constructs.

The self-worship continues, "It's possible that we are merely the first in a great wave of species that will take up tool-making and language. But it's also possible that intelligence just isn't one of natural selection's preferred modules. We might think of ourselves as nature's pinnacle, the inevitable endpoint of evolution, but beings like us could be too rare to ever encounter one another. Or we could be the ultimate cosmic outliers, lone minds in a Universe that stretches to infinity."

This is just the same narcissistic insanity as the astronomer who said we needed to explore Mars "to answer that most important question: are we all alone?" as this culture destroys life on this planet.

The exact same insanity shared by nearly everyone in this culture.

It's extraordinary to me that people who say that evolution is based on random mutations can at the same time say that we are the pinnacle. If the mutations are random, there is no pinnacle. If the mutations are random, there is no inevitable endpoint. Although in the case of this planet, human supremacists are guaranteeing that we are the endpoint by stopping all evolution. And please note, yet again, how all through this he is naturalizing the destruction of the planet.

It's all based on the unquestioned belief that we are more intelligent than nonhumans. It is based on the unquestioned belief that we are superior to nonhumans. It is based on the unquestioned belief that intelligence is manifested by destructiveness.

In the rest of the interview Musk talks about his plans to get a million people to Mars in the next hundred years on 10,000 space ships (presumably built by his company at taxpayer expense), shooting these transports into space at rates "that would convert Earth's launch pads into machine guns, capable of firing streams of spacecraft at deep space destinations such as Mars." He talks about mining Mars, like this culture has mined the earth. He talks about how the entire universe might be a giant computer simulation.

It's all nuts. And it's nearly always couched in terms that are Biblical, magical, childish, or all three. The journalist repeatedly talks about the "sacred mission" of space colonization, and he describes Musk's "cathedral-like rocket factory," then finishes his description by saying, "The place felt something like Santa's workshop as re-imagined by James Cameron."

The Biblical/magical/childish imagery continues, as here we go again with the Noah metaphor: "It's possible to read Musk as a Noah figure, a man obsessed with building a great vessel, one that will safeguard humankind against global catastrophe [except, of course, that only humans will survive, but then again, "Fuck Earth"]. But he seems to see himself as a Moses, someone who makes it possible to pass through the wilderness—the 'empty wastes,' as Kepler put it to Galileo—but never sets foot in the Promised Land."

The article concludes: "He is a revivalist, for those of us who still buy into cosmic manifest destiny. And he can preach. He says we are doomed if we stay here. He says we will suffer fire and brimstone, and even extinction. He says we should go with him, to that darkest and most treacherous of shores. He promises a miracle."

❈ ❈ ❈

All this talk of miracles and preaching and fire and brimstone and sacred missions is not coincidental. The technotopian vision is just a secular version of the same monotheistic conceit that life on Earth is a vale of tears and the real glory is in heaven. It doesn't much matter whether you believe the only meaning comes from a God who looks like an old man with a beard, or the only meaning comes from things created by man, you're still saying that the earth is meaningless. You're still showing contempt and hatred for the earth. And it doesn't much matter whether the God you created tells you that you should have dominion over the earth, and all creatures on earth should fear you, or whether you believe it is human's manifest destiny to convert the earth into machines and pollute the earth (cuz that's what intelligent beings do), and you not only make all creatures fear you, you drive them extinct, you're still destroying the place. It doesn't matter whether you have the God you created tell you that you are the Chosen People (or Chosen Species), or whether your own delusions tell you that your vast intelligence is "a single candle flame, flickering weakly in a vast and drafty void," you still think your chosen stature allows you to exploit and/or exterminate all those you perceive as lesser than you, which is everyone. And it doesn't much matter whether you believe heaven is way up in the stars where God lives, or whether you believe heaven is way up in the stars where you want your space ships to go, you still don't believe that the earth is a good place to live.

There are some differences though. One is that it used to be that at least God was more powerful than Man. Now, though, we've gotten rid of that silly God talk and it is *we* who are on the path to becoming god-like in our capabilities. Another is that in the olden days the Heaven to which the hell on earth was contrasted was at least marginally pleasant, so long as you like harps, and petting zoos that contain both lambs and lions. This new heaven on Mars promised by Musk sounds more like hell: "If you were to stroll onto its surface without a spacesuit, your eyes and skin would peel away like sheets of burning paper, and your blood would turn to steam, killing you within 30 seconds. Even in a suit you'd be vulnerable to cosmic radiation, and dust storms that occasionally coat the entire Martian globe, in clouds of skin-burning particulates, small enough to penetrate the tightest of seams. Never again would you feel the sun and wind on your skin, unmediated. Indeed, you would probably be living underground at first, in a windowless cave, only this time there would be no wild horses to sketch on the ceiling."

It gets even better: "Cabin fever might set in quickly on Mars, and it might be contagious. Quarters would be tight. Governments would be fragile. Reinforcements would be seven months away. Colonies might descend into civil war, anarchy or even cannibalism, given the potential for scarcity. US [sic] colonies from Roanoke to Jamestown suffered similar social breakdowns, in environments that were Edenic by comparison. Some individuals might be able to endure these conditions for decades, or longer, but Musk told me he would need a million people to form a sustainable, genetically diverse civilisation."[141]

So basically the heaven he's promising us is worse than the worst prison cell in the U.S. penal system. At least those in solitary confinement get an hour a day to walk in a walled-in exercise area where they can see the sky and breathe outside air. At least there *is* outside air.

But from the perspective of capitalists—and more broadly, that of an authoritarian technics—this really is heaven. Recall that a central point of agriculture has been to make people dependent on those in power for their food: if you control someone's food, you control their lives, which means you control their labor. The people in Musk's heaven would be dependent on those in charge for the very air they breathe. The God of capitalism/Authoritarianism is smiling.

❀ ❀ ❀

This is the endpoint of human supremacism.

No, the endpoint of human supremacism is what we see around us: the complete insanity of those who suffer the mental illness of human supremacism; and the extermination of all those who don't.

❀ ❀ ❀

I want to mention a few headlines.

"Trees vs. Humans: In California Drought, Nature Gets to Water First."[142] The article argues that because forests naturally retain water (which is a good thing, in terms of forest and river health, and flood, siltation, and landslide prevention, but which this article calls a "sin of nature"), and because California's reservoirs, which are by definition unnatural and are necessary for agriculture, have been depleted by drought (and by agriculture), the solution, brought to you by your friendly corporate foresters, is to cut down the trees. I'm not kidding. And yes, I often wonder at the lack of intelligence in this culture.

A forest activist friend of mine responded to this: "Why don't they just cover the hills in Visqueen and be done with it?"

The next headline: "Texas' Top Toxicologist: EPA's New Smog Regulations Unnecessary, Just Stay Indoors." A relevant quote: "Ozone is an outdoor air pollutant because systems such as air conditioning remove it from indoor air. Since most people spend more than 90 percent of their time indoors, we are rarely exposed to significant levels of ozone."[143]

The next headline (with pull-quote): "Drive to Mine the Deep Sea Raises Concerns Over Impacts: Armed with new high-tech equipment, mining companies are targeting vast areas of the deep ocean for mineral extraction. But with few regulations in place, critics fear such development could threaten seabed ecosystems that scientists say are only now being fully understood."[144] Where have we seen this before, where "development could threaten" natural communities (and inevitably does)? Oh, that's right, it would be everywhere this culture has "developed."

And the next: "Public Forests Sacrificed to the Biomass Industry."[145]

And while Elon Musk feeds at NASA's public trough, another headline: "After 42 Years of Charting the Health of Our Seas, Scientist's Studies Now Face the Axe."[146]

Or this headline: "The Surprising Reason Abandoned US Mines Haven't Been Cleaned Up." The "surprising" reason? "No one really cares."[147]

Recall what Mumford said about the priorities of a culture driven by authoritarian technics: "As with the earliest forms of authoritarian technics, the weight of effort, if one is to judge by national budgets, is toward absolute instruments of destruction, designed for absolutely irrational purposes whose chief by-product would be the mutilation or extermination of ~~the human race~~" life on earth.

The next headline: "Dead Babies Near Oil Drilling Sites Raise Questions for Researchers."[148] Yes, you read that correctly. Babies are dying near oil drilling sites, and the response by this culture is not to stop the murders, but that the murders merely "raise questions for researchers."

Or: "Fracking or Drinking Water? That May Become the Choice." The article begins: "Fracking for oil and natural gas—or having enough water to drink. That's the possible dilemma facing a number of countries including the United States, according to a new report released by the World Resources Institute last week—though experts disagree on the real implications of the report and what should be done about it."[149] Yes, it is perfectly sane to consider the choice between having water to drink and oil and gas from fracking a dilemma, and it is perfectly reasonable for "experts" to disagree as to what should be done about this. Of course since pollution is one of the discernible measures of intelligence, fracking is the only intelligent choice.

And we all see the relationship between those last two articles, right? Although experts may disagree on the implications of this relationship.

Another headline: "We're Damming Up Every Last Big River on Earth. Is That Really a Good Idea?"[150]

Followed by: "Hydropower May Be Huge Source of Methane Emissions."[151]

And only two more. First, "Amazon Rainforest Losing Ability to Regulate Climate, Scientist Warns." It begins, "The Amazon rainforest has degraded to the point where it is losing its ability to benignly regulate weather systems, according to a stark new warning from one of Brazil's leading scientists."[152] The article also talks about the collapse of the forest if the weather changes.

And we all know this culture's response to that emergency, which is given in the final headline: "Amazon Deforestation Picking Up Pace, Satellite Data Reveals: Data indicates 190% rise in land clearance in August and September compared with same period last year."[153]

CHAPTER TWENTY

SELF-AWARENESS

If your emotional abilities aren't in hand, if you don't have self-awareness, if you are not able to manage your distressing emotions, if you can't have empathy and have effective relationships, then no matter how smart you are, you are not going to get very far.

DANIEL GOLEMAN

Many human supremacists love to talk about the "mirror test" of self-awareness, in which you put a mirror in front of some nonhuman to see if the nonhuman recognizes itself, in which case it is declared to be self-aware (though not as self-aware as us, of course!). Very few nonhumans pass this particular test, which is I'm sure one reason the test is so beloved by so many human supremacists. I'm sure it's also a reason this test is sometimes called the "gold standard" of indicating whether some creature is "self-aware."

The test is fraught with problems. First, there's our old friend tautology: humans conceptualized the experiment presuming that humans are self-aware and nonhumans are not, and then devised a test humans can pass and nonhumans cannot. Great job. My understanding of my nonhuman neighbors is so much greater now.

Next, there's our old friend anthropomorphization: the presumption that the self-awareness of others must match the form of our own self-awareness, and further that it must match one specific chosen form of self-awareness. Can there not reasonably be said to be other ways to be self-aware? I know that for myself, I am at least on occasion self-aware even when not looking at a mirror. Imagine that! And I think we can say that humans were probably still self-aware before the invention of the mirror. Or what about the self-awareness of a caterpillar who knows she has a parasite egg in her and that she must eat certain foods or she will die? Do you know when you have parasite eggs in you? If not, then gosh, you must not be very self-aware. Or what about the self-awareness of plants who know how to change the taste of their leaves? Can you change the taste of your own flesh to make yourself less palatable to predators? To this latter you can reply, "Yes, that's why I eat at McDonald's."

And of course there are lots of beings whose primary experience of the world is not visual. How well could you pass a self-awareness test that involves you being able to hear and respond to your own echolocation signals? What? You say you can't hear your own echolocation signals? That's a sure sign of a lack of self-awareness.

For crying out loud, anyone who feels hungry is self-aware, obviously, or they wouldn't know they're hungry. Anyone who attempts in any way to stop pain or discomfort or to continue to receive pleasure is self-aware, or they wouldn't know the state they're trying to change or perpetuate.

Ah, the human supremacists insist, we understand that the tiger is aware of its hunger, but is the tiger aware that it is aware of its hunger? That is the question. To which I ask, are the human supremacists aware of their own hunger? Are they aware of the violation imperative that drives this culture? Are they aware that they've indentured themselves to authoritarian technics and that they are no longer fully human, that they are, to use the Buddhist term, hungry ghosts: undead and unliving spirits of the greedy, "who, as punishment for their mortal vices, have been cursed with an insatiable hunger"?

And then there's the presumption that the behavior of captive animals (or plants) tells us something about either their interior lives or what their personalities, relationships, or lives are like when they're free. The behavior of captive beings tells us about the behavior of imprisoned and (by definition) abused beings.

If you take a lizard from his home, put him in a cage, and present him with a mirror, what the fuck do you want him to do with it?

Let's turn this around and see how you feel about it. You're sitting in your home, minding your own business, when suddenly several unbelievably ugly creatures burst in. They throw a net over you and begin dragging you out the door. Members of your family rush to save you, and the unbelievably ugly creatures kill them with casual swats. You see one member of your family huddling in a corner, making sounds of terror you did not know humans could make. Another casual swat and the sounds stop. The net is hauled outside, and you are put into some sort of container. You feel the container being lifted, and then lifted, and lifted. It takes what seems like hours for you to realize that what you've read about in the tabloids and bad science fiction novels has happened to you: you've been abducted by aliens. The aliens take you to their ship, and over the next days and weeks and endless months they perform tests on you. Do you think your behavior will be the same on their ship as it was in your home, with your family? Do you think your behavior will ever again be the same? And what if these aliens put something in your room, some *thing* you'd never seen before they brought you to this terrible place? Here, in this alien prison, you've seen them preening before it, and making gawdawful faces at it—at least you think those are their faces—and now they're staring at you—at least you think they're staring, and you think those are eyes. You look at this *thing* more closely. They evidently see—perceive is probably a better word, since you don't think those are eyes after all—themselves in it, but frankly their senses must be different than yours, because you don't see what's so great about it. Frankly it's creepy. But then again, so is everything about this place . . .

Because you failed to respond as they wished to this new device, the aliens put into your cage, and the aliens decide—quite rightly, according to their evidence and their belief system—that all you humanbeast-machines (as one of their philosophers puts it) lack self-awareness.

At some point the aliens realize how important vision is to you, and that you see with your eyes. So in order to further their understanding of human behavior, and of course in order to get further grants, they surgically blind you. Sitting in the eternal dark of your cage in some unfathomably huge complex, unimaginably far from your home and from those you love—those who may be still alive among those you love—for some reason you remember an article you read years ago. It was about mice who love to sing, and about what happened to these mice, about how they were put in cages, about what scientists did to them then. Day after day—or at least you think it's day after day, since in your cell and in your own private darkness there is never any natural indication of the passage of time—you obsess about this article. But for the life of you, you can't figure out why it is so important to you.

Before we go to the biggest problem with the mirror test of self-awareness, let's take one more detour, through four stories of this culture's hatred of the natural world. The stories themselves aren't unusual.

A friend in India just told me of a beautiful blue frog who lives a few hundred miles from my friend's home. It is critically endangered; it has only been seen four or five times in the last century. Recently, four of them were seen in one place. These frogs live under rotting logs. When news got out of this group of four frogs, scientists, "naturalists," amphibian specialists, and "nature photographers" began to swarm the region. A person who loves frogs and who lives locally has said that as a consequence, there are no longer any undisturbed rotting logs. Every log has been lifted by humans looking for these rare frogs. When these humans find these rare frogs, they pickle them for collections.

Also in India, a "previously undiscovered" species of lizard was identified. It had been killed on a road. Another swarm of specialists formed, and scoured the area. They found precisely one more lizard of this species. They pickled it.

On the Palouse of eastern Washington used to live a white earthworm who smelled like lilies and grew to more than three feet long. They were abundant prior to this culture's arrival, but the plowing of the Palouse has driven them to near extinction. (Wait! How is that possible? I thought that there was no evidence that agriculture is inherently destructive!) In fact, they were thought to be extinct until 2005, when a scientist digging a hole to sample earthworms cut one in half. Since then, scientists have used probes to send electric pulses through the earth. These pulses shock the worms, and though worms are by nature photophobic—they fear light—they come to the surface in an attempt to get away from the pain. Maybe if they shock them hard enough they can get them to jump through hoops. (Oh, and by the way, wouldn't coming to the surface of the soil to avoid the shocks even though they are photophobic be worms doing something "against their nature," and thus be a sign of their intelligence? Oops.) The scientists then use trowels to dig up any worms who resisted their attempts at electrical persuasion. Through this method, scientists have found a few more of these extremely rare worms in the past ten years. And what have the scientists done with these extremely rare worms? Of course, collected them to put into labs.[154]

And one final story for now. Recently, the body of a man was discovered in California. He had lived off the grid in a forest and died of a heart attack. A black bear found his dead body, dragged it to the bear's home, and ate him. No big deal, really. It's what happens to dead bodies. It's what I hope happens to mine. But here are some consecutive sentences from the article: "'The bear does not pose a public threat,' the paper quoted Andrew Hughan, spokesman for the California Department of Fish and Wildlife, as saying. 'It was just doing what bears do.' Officials tried to trap and kill the bear but called off their attempt because it appeared doubtful the bear was still in the area of the man's home in Redway."[155]

What? Wait. What? The bear doesn't pose a public threat, but then without comment the next sentence says that they wanted to trap and kill it? Why?

Oh, that's right, humans are special, even after we're dead. By which I mean, humans are not part of nature, even after we're dead. By which I mean, humans are not supposed to give back to the earth, even after we're dead.

And at last to the biggest problem with the mirror test of self-awareness, which is that I find it both extraordinary and all-too-expected that members of this culture have the gall to look down on *anyone* as lacking self-awareness. Most humans in this culture—particularly human supremacists, or rather supremacists of any sort—fail the mirror self-awareness test spectacularly. Oh sure, most of us can use a mirror well enough to comb our hair or make sure we don't have boogers hanging out of our noses, and most of us can recognize ourselves well enough in the mirror to become anxious about our looks, but I don't think that an ability to use a mirror to comb one's hair necessarily implies self-awareness on any sort of significant level.

Especially when you're killing the planet.

When we look in the mirror, what do we see?

We see God's image on Earth or the pinnacle of evolution. We see the greatest gift the universe has ever given itself. We see the bringers of the light of consciousness to the universe. We see the universe knowing itself.

We see those whose responsibility it is to bring this light of consciousness everywhere. When we look at our technics, we see only our own brilliance.

When others look at us, however, they see something completely different. They see those who have become Death, destroyer of worlds. They see those who invent machines to outsource Death, and to outsource and facilitate the destruction of worlds. They see those who lack the self-awareness to perceive, much less comprehend, that they have become Death, destroyer of worlds. They see those who lack the perceptiveness or honesty to acknowledge that they are murdering—sorry, reorganizing—the planet. They see those who are so entranced by the technics that control them that they believe there is "no evidence" these technics are inherently destructive, and that there are no "costs" associated with these technics.

They see beings who care more about money than life.

They see beings who care more about power than life.

They see beings whose imagination is so impoverished that they cannot imagine living without industrially-generated electricity.

And they see beings whose empathy is so impoverished that they *can* imagine living without salmon, passenger pigeons, whales, snub-nosed sea snakes, ploughshares tortoises, and on and on.

They see those who when they even acknowledge the Death they cause—in their agriculture, in their economics, in their science, in their religions and philosophies, in the extinctions they cause—they see only how this Death will affect them and the technics they serve.

When others look at us, they see those who have so destroyed their own empathy that they don't even acknowledge—can no longer even conceptualize—that anyone else actually subjectively exists. It is impossible to be less empathetic than that. They see those who have so destroyed their own empathy that they routinely torture those they perceive as below them on the insane Great Chain of Being, that hierarchy they had the lack of empathy and creativity to come up with in the first place. They see those who have so destroyed their own empathy that the males of the species now routinely rape the females of the species. They see those who have so destroyed their own empathy that they attempt to destroy the empathy of those unfortunate enough to encounter them. They see those who have so destroyed their own empathy that they have developed an economics, a politics, a science, an epistemology—an entire worldview—based on projecting this lack of empathy onto the real world, a worldview that makes a virtue and a fetish of this lack of empathy, that attempts to naturalize this lack of empathy, that attempts to pretend empathy doesn't exist in the real world. They see those who have so destroyed their own empathy that they use the empathy of others—empathy they are all the while pretending does not exist—to kill these others. Recall the whalers who would intentionally wound, but not kill one whale, then kill all others who came to help. Recall those who would do the same to the Carolina parakeets. They drove Carolina parakeets extinct. They are driving the world extinct.

When others besides human supremacists look at us, they see the worst thing that has ever happened to this planet.

When we look in the mirror we see the only creature who is fully intelligent, with a brain that is the "most complex phenomena in the universe."

When others look at us they see those who are stupid enough to put poisons on our own food, to poison our own drinking water. Those who are stupid enough to murder—sorry, reorganize—the planet that is our only home.

When we look in the mirror we see the only creature who is fully imbued with the ability to make choices.

If this is the case, and if actions speak louder than words, then we are evidently choosing to kill the planet.

R.D. Laing wrote, "At this moment in history, we are all caught in the hell of frenetic passivity. We find ourselves threatened by extermination that . . . no one wishes, that everyone fears, that may just happen to us 'because' no one knows how to stop it. There is one possibility of doing so if we can understand the structure of this alienation of ourselves from our experience, our experience from our deeds, our deeds from human authorship. Everyone will be carrying out orders. Where do they come from? Always from elsewhere. Is it still possible to reconstitute our destiny out of this hellish and inhuman fatality?"[156]

So, when others see us they see those who have enslaved themselves to their own creations, who are unable or unwilling to question these creations even when these creations are killing the entire planet. They see those who at one time had the ability to choose, but long ago surrendered that ability in exchange for the ability to leverage power and outsource killing.

Choices? Choices? We don't need no stinking choices.

We just follow wherever the technics lead.

When we look in the mirror we see the only source of meaning in the entire universe.

When others look at us they see destroyers of meaning, converters of forests to parking lots, prairies to monocultures, rivers to the industrial electricity without which we can't imagine life. They see us as the destroyers of all complexity, the great simplifiers, making things simple so our simple minds can (still fail to) understand them.

When we look in the mirror we see ourselves as the creators of great art.

When others look at us they see the destroyers of art, the destroyers of beauty, the destroyers of bison and blue whales and monarch butterflies and old growth forests and prairies at dawn and oceans full of fish. What is more beautiful, the sound of a meadowlark or the sound of a highway? The sight of a river or a dam? The smell of a forest or a city? If you are in a city, look around: once, this place, too, was wild and beautiful.

It was written of passenger pigeons: "I have seen them move in one unbroken column for hours across the sky, like some great river, ever varying in hue; and as the mighty stream sweeping on at sixty miles an hour, reached some deep valley, it would pour its living mass headlong down hundreds of feet, sounding as though a whirlwind was abroad in the land. I have stood by the grandest waterfall of America and regarded the descending torrents in wonder and astonishment, yet never have my astonishment, wonder, and admiration been so stirred as when I have witnessed these birds drop from their course like meteors from heaven."[157]

Gone, all gone. Killed by this culture that is Death, destroyer of worlds.

Or this again, also about passenger pigeons: "Every afternoon [the pigeons] came sweeping across the lawn, positively in clouds, and with a swiftness and softness of winged motion, more beautiful than anything of the kind I ever knew. Had I been a musician, such as Mendelssohn, I felt that I could have improvised a music quite peculiar, from the sound they made, which should have indicated all the beauty over which their wings bore them."[158]

And, once again, all gone. By this culture that devours beauty just as it devours land.

I recently watched a documentary on the US invasions of Iraq. There were lots of photos of tanks and trucks and troops moving through the countryside. What impressed me most were the desert backdrops. You could look from horizon to horizon and not see a single plant.

Before this culture, that was cedar forest so thick that sunlight never touched the ground.

We have become Death, destroyer of worlds. We are driven by our insane—and insatiable, because impossible—quest for validation of our self-perceived superiority. We are driven to destroy all that is alive and free and beautiful and wondrous and meaningful *and is not made by or dependent upon us, not under our control.*

I've never forgotten the line I read so many years ago: If animals could conceive of the devil, his image would be man's.

They can, and I'm sure they do.

Our failure at the mirror test of self-awareness reminds me of nothing so much as Oscar Wilde's *The Picture of Dorian Gray*, the central conceit of which is that as the main character becomes increasingly vile, his countenance remains clear, but a portrait of him changes to reflect who he has become. When we look in the mirror, we continue to see a bright and beautiful and intelligent and wonderful being, but who we actually are has become dull and ugly and stupid and as vile as it is possible to be.

And we can't see a fucking thing. We can say, with a clean (because completely eradicated) conscience, "I see no evidence of any inherent destructiveness in what we do or who we have become."

CHAPTER TWENTY-ONE

"REBOOTING THE WORLD," or
THE DESTRUCTION OF ALL THAT IS

I will wipe mankind, whom I have created, from the face of the Earth—men and animals, and creatures that move along the ground, and birds of the air—for I am grieved that I have made them.

GENESIS 6:7

I propose that the core of sadism, common to all its manifestations, is the passion to have absolute control over a living being, whether an animal, a child, a man, or a woman.

ERICH FROMM

The world of life has become a world of 'no-life'; persons have become 'nonpersons,' a world of death. Death is no longer symbolically expressed by unpleasant-smelling feces or corpses. Its symbols are now clean, shining machines; men are not attracted to smelly toilets, but to structures of aluminum and glass. But the reality behind this antiseptic façade becomes increasingly visible. Man, in the name of progress, is transforming the world into a stinking and poisonous place (and this is not symbolic). He pollutes the air, the water, the soil, the animals—and himself. He is doing this to a degree that has made it doubtful whether the earth will still be livable within a hundred years from now. He knows the facts, but in spite of many protesters, those in charge go on in the pursuit of technical 'progress' and are willing to sacrifice all life in the worship of their idol. In earlier times men also sacrificed

their children or war prisoners, but never before in history has
man been willing to sacrifice all life to the Moloch—his own and
that of all his descendants. It makes little difference whether he
does it intentionally or not. If he had no knowledge of the possible
danger, he might be acquitted from responsibility. But it is the
necrophilious element in his character that prevents him from
making use of the knowledge he has.

ERICH FROMM

You feel the last bit of breath leaving their body. You're looking
into their eyes. A person in that situation is God!

SERIAL SEX KILLER TED BUNDY

Before we start to wind down, I want to tell you three more stories about our addiction to authoritarian technics. I've written elsewhere how the word *addiction* comes from a root that means "to enslave," in that a judge would issue an *edict* enslaving someone. To be an addict is to be a slave, in this case to the authoritarian technics.

The first story is that I recently saw a TV advertisement for an automobile in which an actor states, "Our species is defined by the tools we use. That's how we got to the top of the food chain." As the actor is saying this, we see a big Dodge Penis—I mean, Caravan—drive by. The point seems to be that you can get to the "top of the food chain," which, I suppose in this case, means at the top of this culture's hierarchy, simply by having a big penis—I mean, tool, I mean, automobile. The thing I find interesting is that the ad tells a lie and tells a truth. The lie is that the hierarchy he describes—whether you call it the Great Chain of Being or the "food chain"—exists at all. There is no top of any food chain. It's all cycles within cycles. You eat the fish who ate the worm, and in time the worm eats you. It doesn't matter whether you are a gnat or an elephant, you eat and you will be eaten. That's life. Get over it.

But there is a germ of truth in the statement that our tools define us. I'm thinking especially of the second definition of the word *define*, which is "to fix or mark the limits of: *demarcate*, as in 'rigidly defined property lines.'" It comes from the root *de-finire*, to limit, end, from *finis*, boundary, end. In this sense, our tool usage does define us. It limits us—as in, to provide one example among far too many, destroying our imagination such that we can no longer imagine living without industrial electricity, even as its generation kills the planet we need to survive—and also, as should be clear by now, our tool usage in this culture threatens to end us, as well as almost everyone else. So sure, our culture—not our species—is defined—that is, limited and ended by—the tools we use.

The next story also involves television. I was flipping through the channels at my mom's, and came across a program—and isn't that a wonderful use of that word?—on the History Channel entitled *101 Gadgets That Changed the World*. It was for the most part the same sort of narcissistic pablum as in *The Atlantic* article I deconstructed earlier, only this one focused on self-congratulatory buffoonery instead of the more overt worship of authoritarian technics. In other words, instead of extolling instruments to facilitate slavery and outsource death, this show focused on gadgets like duct tape, sunglasses, derringers (!), floppy discs, and MP3 players. I mention it not only because while I think duct tape is pretty handy and floppy discs were kind of cool, it is a measure of our enslavement to machines—and our allowing ourselves to be defined by the tools we use—to call these "world-changing." The world is much bigger than my unsuccessful attempts to use duct tape to repair my garden hose that got chewed on by a bear, or my rather more successful attempts to install *Wolfenstein 3D* on my computer circa 1990. I mention it more because of gadget number ten, which was the lightbulb. Sure, lightbulbs have changed the world, in that now they can collectively be seen from outer space, and because they have allowed us to stay up reading all night without having to go the Abraham Lincoln route of reading by the dying embers, but lightbulbs, just like pesticides, just like automobiles, just like every other tool that defines us, reveals as always our blind spot when it comes to the downsides of technologies.

What? There's a downside to lightbulbs? There is "no evidence" of a downside to lightbulbs, just as there is "no cost" to lightbulbs. Never mind the ecological harm caused by their manufacture, transportation, use, and disposal. Never mind the hundred million migratory songbirds killed by them each year by flying into lit skyscrapers, and never mind the uncountable insects killed by them.

Let's leave those aside. They don't count to most people. But here's my point: within twenty years of the invention of lightbulbs, night shifts at factories had become commonplace, and consumerism had tripled. From the perspective of capitalists, this is a good thing. From the perspective of the Magnificent Bribe, this is a good thing. From the perspective of the world, and from the perspective of our humanity, not such a good thing.

Even a gadget has consequences.

I had a friend who thought the lightbulb was the single worst human invention, because of what it did to our sleep. I think our pineal glands would agree with him. I think sleeping pill manufacturers would disagree.

All of my life I have suffered from extreme and intractable insomnia, I think in part because of the abuse I suffered as a child, which led to nightmares, night terrors, and almost as many sleepless nights as not. A decade of therapy and a year writing *A Language Older Than Words*, which helped me make meaning of the suffering and helped me find what many trauma experts would call a "survivor's mission," got rid of most of the nightmares and night terrors, but the insomnia remains.

That said, I don't think lightbulbs have done me any favors. I think this because years ago, when I lived in Spokane, Washington, an ice storm took out electricity to my part of town for a couple of weeks. I had a woodstove and plenty of guilt-free wood (that I got for cheap from a pallet factory that had so many mill-ends they trucked most of them directly to the incinerator, which means the wood I burned was going to be burned anyway; time for short pants even though the windows are iced over!), so heat wasn't a concern. My mom lived not far away and was on city water, so likewise, water wasn't a concern. Spokane is far to the north, so sundown was around four, and sunrise around eight. Consequently I went to bed around six, and fell asleep each night around seven or eight. I'd wake up at four in the morning, look at the iridescent hands on my travel alarm clock (the alarm clock was number nine of the world-changing gadgets, because the nightstand version of it "helped drag the Industrial Revolution out of bed"), and delight in the fact that I had four more hours to sleep before dawn. After maybe ten days of this, I was, for the only time in my adult life, completely refreshed.

Then the electricity came back on, and my addiction to electric light bulbs (and reading till midnight) kicked right back in, and with it, my light-induced insomnia.

I've noticed that when I go camping, my insomnia disappears after a night or two. I've read that I'm not alone in this.

When I talk about this culture's addictions to these technologies, I include myself.

And please don't use my own addiction—as I know a lot of lifestylists will—as an excuse to dismiss my larger analysis. First, the honest reflections of a heroin addict might have more credibility when speaking of that addiction than might that of a non-user. Second, the first step toward recovery from addiction is to admit there is a problem, and this entire book (in fact my entire life's work) is aimed toward getting us as a culture to admit we are addicted to this terribly destructive way of life, because if we don't acknowledge that these addictions even exist, we have no hope of breaking them. And third, the point here is not and has never been purity, and while removing the lightbulbs from my own home would help me sleep, it wouldn't do a fucking thing to help the migratory songbirds or the insects, and it wouldn't do a thing to stop consumer culture or any of the other costs of this particular technics. There are no personal solutions to social problems.

Indeed, I think the story of my temporary escape from that addiction through the removal of its source points to one of the few realistic ways to get past these addictions to authoritarian technics, get past them to a better way of life. You know what that way is.

The third story has to do with a recent cover article in *Newsweek* about geoengineering, titled, "Science to the Rescue: Rebooting the Planet." Yes, one should never anthropomorphize, except when one is projecting machine/computer language onto the natural world. And by this point in the book, do I really have to point out that "rebooting" the planet has been precisely what this culture has been aiming for since the beginnings of human supremacism? The point from the beginning has been to "shut down" the natural world, in other words kill it, and then use our own technics to "restart" some facsimile of it. It's the story of Noah's Ark. It's the story of the Second Coming of Jesus, with the destruction of the earth and its replacement by heaven (or in the new version, technotopia).

It's the story of cities (wiping out all native life and then converting the land solely to human use). It's the story of agriculture (wiping out all native life and then converting the land solely to human use). It's the story of pesticides. It's the story of genetic modification. It's the story of scientific experiments (wiping out all variables but one in a laboratory, and then manipulating that last variable in order to, as Dawkins put it, make matter and energy jump through hoops on command; or, as Descartes put it, to torture nature into revealing her secrets). It's the management story. It's the neo-environmentalist fantasy of a world controlled by us where "anything goes." It's the standard abuse story, where the perpetrator breaks down and remakes the victim. It's the endpoint of this whole machine culture.

And do I have to mention that when they say "Science to the rescue," they don't actually mean the rescue of the planet: they mean rescuing this culture from the effects of "turning off"—sorry, reorganizing; oh fuck it, murdering—the world?

Of course they don't mean the rescue of the planet. The machine über alles.

Early on, the article lays out what's going on, "With the Earth warming at a rate 10 times faster than the heat-up after the last ice age, scientists are looking at anything they can use to stop climate change."

The problem is that when someone in this culture says they're looking at "anything they can use to stop climate change," they really mean looking at "anything they can use" *except* the sense God gave a goose, and then using that God- or evolution- or nature-given sense to question authoritarian technics, to question human supremacism. That is, they really mean they're looking at everything except the things that matter most.

You could—and frankly, a lot of human supremacists do—argue that questioning authoritarian technics—which means questioning everything civilization is based on, including agriculture, including human supremacism—is insane and monstrous.

I think that when what is at stake is life on this planet, and when it's plain to see that from the beginning this way of life has been functionally destructive, that *not* questioning this way of life is what is insane and monstrous. What's insane and monstrous is preferring this way of life over life on earth.

So, in the article, what *is* meant by scientists "looking at anything they can use to stop climate change"?

Well, some of the options actually make sense. The author states, "It's not crazy to think humans could come up with ways to change the makeup of the planet; after all, humans have already reengineered the earth by accident [sic]. Across the planet we've torn down carbon-capturing forests to make room for farms, so we could feed our growing populations. And David Edwards, a professor of conservation science at the University of Sheffield, is starting to think that one of the best ways to geoengineer the planet is to figure out a way to bring those forests back."

Actually, it's really easy to bring the forests back: stop destroying them and let them come back. It's what they want to do, and it's what they do best.

But stopping deforestation and encouraging reforestation becomes a problem when you're living in a culture with an extractive economy. When your economy requires and rewards deforestation, and you don't want to destroy your economy—if, in fact, you can't even *question* your economy—then it becomes necessary for you to try to "figure out a way to bring those forests back." It also becomes, on the largest scale, impossible, for the reasons I laid out earlier; this is all just the environmental version of the anti-empire activists who still want the goodies of empire. These people still want the goodies that come with an economy based on drawdown, and hope to find a way to get them without, well, drawdown. It's impossible to have overshoot without having the effects of overshoot.

Consider the dead zones in the oceans, which are primarily caused by agricultural runoff (high fertilizer concentrations cause algae populations to explode, then crash; their decomposition depletes oxygen in the water, and oxygen-breathing beings die). Just today I was talking with someone who works on issues associated with dead zones. He said that of the more than 400 of these dead zones across the world, only one has disappeared: the one in the Black Sea.

I asked him what's different about that one.

"The dead zone went away because the collapse of the Soviet empire caused the collapse of the region's economies, which caused chemical fertilizers to become too expensive to use." He paused, then continued, "Now I'm not saying we need to end empires . . ."

I said, "I'll say it."

There's a cause-and-effect relationship between destructive activities and the destruction they cause. There is a cause-and-effect relationship between *not stopping* destructive activities, and *not stopping* the destruction those destructive activities cause. So many people want the destruction to cease, but don't want to stop the destructive activities. And that, of course, is the main point of geoengineering. That's the main point of most of what passes for environmentalism as well.

I don't understand why more people don't understand this.

You wanna stop global warming? Well, stop industrial culture. And what happens then? The forests and grasslands and marshes start doing one of the many things they are good at: sequestering carbon.

But of course, the *Newsweek* article didn't mention stopping industrial culture and just letting forests and prairies (and wetlands and coral reefs and seagrass beds) come back. When people in this culture say scientists are considering "everything" to stop global warming, that's never what they mean.

They mean things like dumping iron into the ocean to stimulate blooms of phytoplankton, who will absorb carbon dioxide, then when they die sink to the bottom of the ocean, carrying the CO_2 with them. If done over a great stretch of the Antarctic Ocean, where, if you recall, a reduction in sperm whale numbers already meant a reduction of available iron, this could theoretically sequester almost one-fourth of the carbon dioxide emitted each year by authoritarian technics (i.e., this culture).

Proponents tell us that nothing could go wrong.

That's what they always tell us.

They are always wrong.

One reason they're always wrong is that they lie.

Another is that they see what they want to see, and they don't see what they don't want to see.

Another is that so often they don't particularly care about harm to others. They are socially rewarded for not caring about harm to others.

The most important is that the world is more complex than they're capable of thinking, and more interrelated than they're capable of thinking, and so actions will inevitably have far more consequences than the original actors are capable of conceptualizing, much less predicting.

Given the (essentially zero percent) success rate of human supremacists when they say that nothing will go wrong when they try to manage the world, do you want to gamble life on earth on their say-so?

But the primary point is never really to solve the stated problem anyway, no matter what the human supremacists say. The point is to exert control. The point is to be God. As the geoengineering proponent Richard Odingo said, "If we could experiment with the atmosphere and literally play God, it's very tempting to a scientist."[159]

Actually, that's not a temptation: that's the point. From the very beginning the point has been to destroy the wild nature they fear and hate, and replace it with what they can attempt to control.

So if playing God over an entire ocean isn't good enough for you, we could also, as the *Newsweek* article discusses, send "a fleet of planes into the sky" to spray "the atmosphere with sulfate-based aerosols" that would block sunlight from reaching the earth.

Gosh, what could possibly go wrong?

At least *Newsweek* didn't support some of the craziest ideas, like changing the earth's orbit or putting up thousands of mirrors in space.

The *Newsweek* article ends by turning the focus inward. It states, "Most climate scientists still argue that instead of relying on untested attempts to remake the natural world we've unmade, humans might want to take a look at themselves." Uh, yeah. And especially take a look at our addiction to authoritarian technics. But oops, that's not what the writer meant when she said we should look at ourselves.

Because stopping the murder of the planet would take "a seismic shift in what has become a global value system," instead, she and the scientists suggest "a reimagining of what it means to be human. In a paper released in 2012, S. Matthew Liao, a philosopher and ethicist at New York University, and some colleagues proposed a series of human-engineering projects that could make our very existence less damaging to the Earth. Among the proposals were a patch you can put on your skin that would make you averse to the flavor of meat (cattle farms are a notorious producer of the greenhouse gas methane) [how about a patch to make men averse to rape; or a patch making us averse to all agricultural products; or a patch making us averse to thinking we are the only sentient beings on the planet], genetic engineering in utero to make humans grow shorter (smaller people means fewer resources used) [Yes! That's the ticket! I've always thought the biggest problem with 10,000-ton draglines used in open pit mining is that the operator's cabin is built to hold a six foot human; if we make the human only three feet tall, that will solve the whole problem!], technological reengineering of our eyeballs to make us better at seeing at night (better night vision means lower energy consumption) [Jesus Christ, do you realize what a tiny percentage of industrial electricity is used for lightbulbs so we can read at night?], and the extremely simple plan of educating more women (the higher a woman's education the fewer children she is likely to have, and fewer children means less human impact on the globe). [Finally, one I can agree with, but wouldn't it also be good to educate the men to make it so they don't want to control the women? Actually, skip that: let's just make a fucking patch for the men.]"

The article concludes: "It might be uncomfortable for humans to imagine intentionally getting smaller over generations or changing their physiology to become averse to meat, but why should seeding the sky with aerosols be any more acceptable?[160] In the end, these are all actions we would enact only in worst-case scenarios. And when we're facing the possible devastation of all mankind [sic], perhaps a little humanity-wide night vision won't seem so dramatic."[161]

When faced with the "possible devastation of all mankind [sic]," *Newsweek* proposes everything from manipulating oceans to manipulating the atmosphere to manipulating humans (as one critic of the article put it, turning us into hobbits). But as always, what is left off the table is our addiction to—our enslavement to—authoritarian technics. What is left off the table is any questioning of our human supremacism. What is left off the table is humility.

In this perspective, it is more feasible to engineer the entire planet (or to engineer human physicality) than it is to change this culture's "value system."

Think about that.

This is why human supremacists keep trying to manage the planet, even though each time they do they destroy the biome they are trying to manage: it doesn't really matter to them whom or what they destroy, so long as they keep their way of life going, so long as they get to maintain the illusion of their own superiority, and so long as they get to maintain their "value system." Their "value system" is more important to them than the life of the planet upon whom even their own lives depend. And they call themselves smart?

Unquestioned beliefs are the real authorities of any culture.

And here's the real problem: it's not only mainstream journalists and scientists whose responses to global warming are made absurd by their refusal to question human supremacism and an enslavement to authoritarian technics. Even the writer who complained about turning humans into hobbits responded to the article, "We are already facing the devastation of all mankind [sic]. And science has already provided the means of our 'rescue,' the means of reducing 'the burden humans put on the planet'—the myriad carbon-free energy technologies that reduce greenhouse gas emissions. Perhaps LED lighting would make a slightly more practical strategy than reengineering our eyeballs, though perhaps not one dramatic enough to inspire one of your cover stories."[162]

What?

I honestly don't know which I find more disturbing and surreal: the fact that *Newsweek* seems to think turning us into hobbits is some sort of solution; or the fact that the climate activist critic thinks LED lighting will solve the problem.

Ah, yes, that's right: the problem with draglines isn't that the operator's cabin needs to be smaller! It's that the headlights need to be LED! How silly of me!

For crying out loud, lighting for residential and commercial uses accounts for only about 12 percent of US consumption. Heck, it only accounts for 14 percent of residential use.

Further, every time this culture invents some way to become more energy efficient, the culture doesn't use less energy, but uses that energy efficiency to further ramp up the economy, to produce more saleable stuff; in other words, to convert more of the living to the dead.[163]

It's all insane.

We will go to any length, promote any absurd solution—change the planet, change what it means to be human—in order to avoid looking at the real problem.

This is the power of unquestioned beliefs.

❊ ❊ ❊

Of course we have long since "reorganized" the planet, just as we have long since "reorganized" what it means to be human.

Or more precisely, we have in the service of authoritarian technics "reorganized" the planet, and "reorganized" humanity.

We need to restore them both.

To do so we need to reject authoritarian technics, and we need to reject what "humanity" has become, and we need to reject human suprema-cism. We need to reject supremacism altogether.

❊ ❊ ❊

In addition to this being a book about human supremacism, it's a book about supremacism in general. And ultimately, it's a book about an ideo-logical and physical war that has been going on for ten thousand years between those who hold supremacist and non-supremacist worldviews. The winner of this war determines whether the planet survives. And of course, right now the supremacist side is winning.

The supremacist side in this war believes that members of "our" cat-egory—whatever that category may be—are superior to all others, and that this superiority entitles us to exploit them. In fact, our exploitation of these others is ultimately the primary way we know we're superior. This side believes that difference leads to hierarchy. Men over women. Whites over non-whites. Civilized over indigenous. Humans over non-humans. Animals over plants. Plants over rocks. Mind over matter. Those higher on the Great Chain of Being over those lower. This side in this war believes all life is war, and that the point of life is to defeat others in this war, to scratch and claw and bite, and then to stab and shoot and bomb and poison your way to the top of the hierarchy you've set up (the hier-archy where you already see yourself at the top); and then from the top to exploit all those below you, not merely so you gain the benefits from being so marvelous, but to maintain your position "at the top of the food chain." You and your SUV.

The non-supremacist side in this war believes that difference leads to complexity and community. A forest wouldn't be a forest without the contributions of everyone who lives there. It recognizes that the exploitation of some other is no validation of superiority, but merely the exploitation of some other. It believes that life is not a war, but rather simply life, and the point of life is to live and die, and to do so in such a way that you contribute to the overall health of the community.

The worldviews are simply that: worldviews. They're not reality. Reality is more complex than any worldview. These worldviews have consequences for reality, of course. But they are worldviews nonetheless.

And these worldviews are based on premises. Ultimately, I cannot convince a human supremacist that stones are sentient, any more than he could convince me they're not. I cannot convince him that bears or frogs or caterpillars are sentient, and he cannot convince me they are not. Because we can never know the experience of another.

And because believing is seeing—by which I mean that preexisting prejudice is often more important to us than physical evidence—no matter what evidence I provide for the sentience of nonhumans, a human supremacist can always, as we've seen so many times in this book (and more broadly in the culture) simply ignore the evidence or define or describe away their behavior mechanistically, and continue on his lonely, hierarchical way. Likewise, he could tell me that rocks don't move, and I could respond, "Movement is necessary for sentience? That's awfully anthropomorphic." He could tell me that rocks have no neurons, and I could respond, "Neurons are necessary for sentience? That's awfully anthropomorphic." There is no evidence he could give that would convince me, in part because it is impossible to prove a negative. We could both keep playing our respective games all day, till each wanders away muttering, him that I'm an unscientific idiot and me that he's a bigoted moron.

Or he could say, "Let's put aside our prejudices and really look at evidence."

I say, "Great!"

He says, "You have to admit it's pretty smart to be able to design a rocket that will take us to Mars."

I say, "Good point. And you have to admit there's nothing any creature could do that would be more stupid than to kill the planet."

He says, "Good point. But we invented automobiles."

I say, "Good point. But we're killing the planet."

He says, "We invented modern medicine."

"Killing the planet."

"Computers."

"Killing the planet."

"Nanotechnology."

"Killing the planet."

He says the creation of these technologies trumps everything.

I say killing the planet trumps everything. It doesn't matter what goodies we create if we destroy life on earth. And it certainly doesn't make us smart. Killing the planet is the stupidest thing any creature could do. It trumps every other action.

So then we walk away, him *still* muttering, "Unscientific idiot," and me *still* muttering, "Bigoted moron."

❊ ❊ ❊

And that, really, is the essence of this book, and the essence of the ideological and physical war on which the future of life on this planet rests.

If your definition of superiority means that you are stealing from everyone else on the planet to make "comforts or elegancies" for yourself and the few generations who follow, leaving behind an impoverished or murdered world, then human supremacists really are superior.

If, on the other hand, your definition of superiority has to do with leaving the earth in better health than when you entered—if you value relationships more than power—then this culture fails completely.

❊ ❊ ❊

I'm not suggesting with this book that we shut off our imagination, creativity, or ingenuity. I'm not the one who said nobody could imagine living without electricity. I'm merely making the rather startling suggestion that we use our imagination, creativity, and ingenuity to serve the continuation of life, not use them to serve the concentration of power and the destructiveness that is leading to the end of life on this planet. And if you can't figure out how to use your imagination, creativity, and ingenuity to serve the greater cause of life—in order to leave the world a better place because you were born—well, don't blame me for pointing out that your self-perceived superiority is nothing more than a justification for your exploitation of those around you. And if you can't figure out how to use your imagination, creativity, and ingenuity to make it so the world is a better place because you were born, surely you must understand why I can't take seriously your claim of human superiority. To not be able to make it so the world is a better place because you were born does not seem very imaginative, creative, ingenious, or smart to me. And surely you can see why, if you are grievously wounding the world that is our only home, those of us who care about life would stop you, right?

* * *

Scientists love to talk about how part of the way we learn things is by creating a theory or model, and then seeing how the application of that theory or model plays out in the real world (or at least in a laboratory). If the results are what you originally thought they would be, your theory or model may be correct.[164]

So they can come up with the theory that if they deafen mice who love to sing, these mice might no longer be able to sing. So they deafen the mice, and whaddya know, the mice can no longer sing. The theory is right! They're all geniuses, and humans are superior! Oh, actually this is too small a sample size to be sure. Can we have some more grant money so we can deafen more mice?

On the other hand, if the results don't match your model, the model may be wrong. So I could hypothesize that humans can live on Lucky Charms alone. If someone eats only Lucky Charms and eventually gets sick, I can presume my hypothesis is wrong.

Or here's a hypothesis: the best way to improve a relationship with a lover is to comment scathingly on this person's body each time she or he is naked. Try this out in the real world, and after seventeen exes (fourteen of whom left immediately after the comments but before any further sexual contact, with the other three sticking around only long enough to make certain they hadn't misheard you) followed (after word got around) by thirteen years of celibacy, you might consider your hypothesis to have been effectively shown to be false.

This notion of verifying or falsifying a thesis provides a powerful argument against human supremacism.

How?

Well, for a while now the dominant culture has been acting on the hypothesis that natural selection is based on competition and hyperexploitation; and that just as the capitalists say, each being acting selfishly will (somehow) lead to a successful and resilient community. Result: the world is being murdered. Natural communities are falling apart.

Let's go further back: for several thousand years the dominant culture has been acting on the hypothesis that the world was created for us, that we should go forth and multiply, and we were given dominion over the earth, and that all creatures should fear us. Result: the world is being murdered. Natural communities are falling apart.

Hypothesis: humans are superior to nonhumans. Nonhumans are not as smart as we are. The world consists of resources to be exploited. Result of actions based on this hypothesis: the world is being murdered. Natural communities are falling apart.

How insensate must we be to not see that these experiments are failing miserably?

Unless, of course, the point has from the beginning been to "reboot" the planet, to wipe it clean and for either God or science to remake it in our own image.

But it's still failing, because without a living planet we will not be here, either.

If space aliens instead of human supremacists were conducting this grand open air experiment, we could see all of this, and see that these aliens are completely insane, but because we've been inculcated into this culture, and because we too have been made insane by this culture, we can't see it.

※ ※ ※

Here's the thing: whether or not stones are actually sentient, whether or not redwood trees are smarter or stupider than humans (or whether, as I think is the point, their intelligence is so vastly different as to be incomparable, and cross-species measures of intelligence are both impossible and at best meaningless (and at worst harmful, as we see, when we use them to buttress pre-existing supremacisms)), whether rivers are simply vessels for water or beings in their own right, these are not the primary questions to ask.

Think about it: the Tolowa lived where I live now for at least 12,500 years, if you believe the myths of science; and they lived here since the beginning of time, if you believe the myths of the Tolowa. And they did not destroy the place. When the Europeans arrived here the place was a paradise. I'm not saying the Tolowa were perfect, any more than anyone else is perfect. I'm saying they were living here sustainably.

The dominant culture has trashed this place, as it trashes every place.

The biggest difference between Western and Indigenous worldviews is that Indigenous humans generally perceive the world as consisting of other beings with whom they can and should enter into respectful relationships, and Westerners generally perceive the world as consisting of resources to be exploited.

The western civilized worldview is unsustainable. A belief in human superiority—and the beliefs that nonhumans aren't fully sentient, that rivers aren't beings, and so on—is not sustainable. The fact that it is unsustainable means it is terminally maladaptive. The fact that it is terminally maladaptive means it is an evolutionary dead end. The fact that it is unsustainable makes clear to me that it is also inaccurate: an accurate perception of one's place in the world and actions based on this perception would seem to me to be more likely to lead to sustainability; while an inaccurate perception of one's place in the world, and actions based on *this* perception, would seem to me to be more likely to lead to unsustainability. As we see.

I don't know why more people don't understand this.

I guess because unquestioned beliefs are the real authorities of any culture.

And I guess because most members of this culture have been inculcated into not caring about life on this planet.

That last sentence alone is enough to damn a belief in human supremacism.

※ ※ ※

But you want humans to be superior at something? I'll give you something at which human supremacists excel. Rationalization.

I mean this in two ways. The first is in the way rationalization is usually meant. Rationalize: "to attempt to explain or justify (one's own or another's behavior or attitude) with logical, plausible reasons, even if these are not true or appropriate." Or, "to think about or describe something (such as bad behavior) in a way that explains it and makes it seem proper, more attractive, etc." Or, "to provide plausible but untrue reasons for conduct." Or, "to attribute (one's actions) to rational and creditable motives without analysis of true and especially unconscious motives."

Sound familiar?

This describes human supremacism, which is an "attempt to explain or justify (one's own or another's behavior or attitude) with logical, plausible reasons, even if these are not true or appropriate." Human supremacism is an attempt to explain or justify atrocious and world-destroying behavior. And then so much of this culture's philosophy, its religion, its economics, its epistemology, its science, are all attempts to explain, justify, and/or facilitate human (and other) supremacisms.

We're really good at rationalizing our behavior.

Remember that Timur the Great was able to rationalize his behavior. Hitler was able to rationalize his behavior. The head of ExxonMobile is able to rationalize his behavior.

We are currently rationalizing the murder of the planet.

But *rationalization* means something else, too, at which human supremacists also excel. In terms of "scientific management," rationalization can be seen as the deliberate elimination of information unnecessary to achieving an immediate task. So the process of frying a hamburger at a fast-food restaurant can be said to have been "rationalized" if all extraneous movements and considerations have been removed.

Another way to say this is that human supremacists excel at figuring out "solutions" to discrete "problems" by ignoring everything but the specific "solution" to the discrete "problem." This is essentially the *point* of the scientific method: you try to eliminate all variables but one. Which is a functional problem with the scientific method, and why science is *func-*

tionally so very good at making matter and energy jump through hoops on command, and simultaneously so very destructive of communities.

Step by step, that's how this culture has built itself up. Step by step, the rest of the world has suffered because of it. Other cultures, other species. The entire world.

A human supremacist sees a river. His factory requires electricity, which means he perceives *himself* as requiring electricity. From his perspective, water flowing to the ocean is serving no "beneficial purpose." So he uses the collective knowledge of this culture to build a dam that generates electricity (actually, he has another problem, which is paying for the dam, and the solution of course is to get taxpayers to pay for it).

So, having had the dam constructed for him, he has "solved' the "problem" of "needing" electricity. Humans and nonhumans who lived on the now-inundated lands above the dam pay the costs. Fish who lived in the river pay the costs. Those who ate the anadromous fish who spawned above the dam pay the costs. Those who lived along the lower banks of what was a free-flowing river pay the costs. Those who lived below the dam who require annual flooding pay the costs. Ocean beaches starved of sediment pay the costs. And on it goes.

Of course, in order to get to the point of building a big dam, other discrete problems had to be solved first, such as inventing concrete and steel, but the same process held for each of these, as "problems" were "solved," with each "solution" leading to consequences to be foisted off onto others. And on it goes.

We can go through that same exercise for every significant invention of this culture, where every brilliant "solution" to every pressing "problem" emerges in part or in whole through ignoring the harm done to others by this "solution."

Pesticides. Automobiles. Agriculture. Cities.

I don't mean to discount this ability. The ability to rationalize has allowed members of this culture to accomplish extraordinary things. But these extraordinary things have come at extraordinary costs.

And to this day the self-styled most intelligent being on the planet, whose brain, they say, is the "most complex phenomena in the universe," is unable or unwilling to perceive the enormity of these costs. For the most part, the best we can hope for from human supremacists is that they think that whatever costs they perceive are (of course) more than made up for by the benefits accruing to themselves and others of their class.

After all, that's only befitting of those who are so high on the Great Chain of Being.

✻ ✻ ✻

You've probably noticed I haven't talked about the origins of human supremacism. Some say it began with the domestication of nonhuman animals, as we came to think of these as our dependent inferiors, as our slaves, our beasts of burden. Some say it began with agriculture, where the entire landbase was converted to human use. Some say the model for human supremacism is male supremacism: women are physically differentiable from men, and some men decided that differentiability meant inferiority, and validated their own superiority by repeatedly violating and controlling women; this model was then applied across racial, cultural, and species differences. Some say human supremacism really got its start with the creation of a monotheistic sky god and the consequent removal of meaning from the material earth.

These questions of origins, while interesting and on some levels important, are not vital to the current discussion. Right now this narcissistic, sociopathic human supremacist culture is killing the planet, and we need to stop it. Asking where it started feels a bit to me like wondering about the childhood traumas of the axe murderer who is tearing apart your loved ones. Sure, it's a discussion to be had, but can we please stop the murderer first?

✻ ✻ ✻

Because human supremacism—like other supremacisms—is not based on fact, but rather on pre-existing bigotry (and the narcissism and tangible self-interest on which all bigotries are based), I don't expect this book will cause many human supremacists to reconsider their supremacism, just as books on male or white supremacism don't generally cause male or white supremacists to reconsider theirs. The book isn't written for them. This book is written to give support to the people—and there are a lot of us—

who are not human supremacists, and who are disgusted with the attitudes and behaviors of the supremacists, who are attempting to stop the supremacists from killing all that lives. It is written for those who are appalled by nonhumans being tortured, displaced, destroyed, exterminated by supremacists in service to authoritarian technics. It is written for those who are tired of the incessant—I would say obsessive—propaganda required to prop up human supremacism. It is written for those who recognize the self-serving stupidity and selective blindness of the supremacist position.

It is written for those who prefer a living planet to authoritarian technics. It is written for those who prefer democratic decision-making processes to authoritarian technics. It is written for those who prefer life to machines.

❀ ❀ ❀

I'm sitting again by the pond. The wind still plays gently among the reeds, plays also with the surface of the water.

This time I do not hear the sound of a family of jays softly talking amongst themselves. This time I hear the sound of chainsaws.

The forests on both sides of where I live are being clearcut. I don't know why. Or rather, on a superficial level I do. The people who "own" both pieces of land had a "problem" they needed to "solve." "Problem"? They needed money. Or they wanted money. Or they craved money. It doesn't matter. "Solution"? Cut the trees and sell them.

Never mind those who live there.

So for weeks now I've been hearing the whine of chainsaws and the screams of trees as they fall. For weeks now I've been feeling the shock waves when the trees hit the ground.

Such is life at the end of the world.

❀ ❀ ❀

We end on the plains of eastern Colorado, where as I write this a friend is trying desperately to protect prairie dogs. A "developer" wants to put in a mall on top of one of the largest extant prairie dog villages along Colorado's Front Range. The village has 3,000 to 8,000 burrows.

Prior to this human supremacist culture moving into the Great Plains, the largest prairie dog community in the world, which was in Texas, covered 25,000 square miles, and was home to perhaps 400 million prairie dogs. The total range for prairie dogs was about 150,000 to 200,000 square miles, and population was well over a billion.

Now, prairie dogs have been reduced to about five percent of their range and two percent of their population.

Yet because yet another rich person wants to build yet another mall (in this economy, with so many empty stores already?), much of this prairie dog community will be poisoned. That community includes the twenty or more other species who live with and depend upon prairie dogs. The prairie dogs (and some others) who are not poisoned will be buried alive by the bulldozers, then covered with concrete. This includes the pregnant females, who prefer not to leave their dens.

If you recall, prairie dogs have complex languages, with words for many threats. They have language to describe hawks, and to describe snakes, and to describe coyotes. They have language to describe a woman wearing a yellow shirt, and different language for a woman wearing a blue shirt. They have had to come up with language to describe a man with a gun.

Do they, I wonder, have language to describe a bulldozer? Do they have language to describe the pregnant females of their community being buried alive?

And do they have language to describe the murderous insatiability of human supremacists? And do others? Do blue whales and the few remaining tigers? Do the last three northern white rhinos, all that's left because some human supremacists believe their horns are aphrodisiacs?[165] Do elephants? Did the black-skinned pink-tusked elephants of China? Did the Mesopotamian elephants? And what about others? What about the disappearing fireflies? What about the dammed and re-dammed and re-dammed Mississippi? What about the once-mighty Columbia? What about the once-free Amazon? Do they have language to describe this murderous insatiability?

❀ ❀ ❀

And perhaps more to the point, do we?

❋ ❋ ❋

By the time you read this, the prairie dogs my friend is fighting to protect will probably be dead, killed so someone can build yet another cathedral to human supremacism. And by the time you read this, yet another dam will have been built on the Mekong, on the upper reaches of the Amazon, on the upper Nile. By the time you read this there will be 7,000 to 10,000 more dams in the world. By the time you read this there will be more dead zones in the oceans. By the time you read this there will be another 100,000 species driven extinct.

And all for what?

To serve authoritarian technics, to serve an obsession to validate and re-validate a self-perceived superiority that is so fragile that each new other we encounter must be violated, and then violated, and then violated, till there is nothing left and we move on to violate another.

This is not the future I want. This is not the future I will accept.

❋ ❋ ❋

What I want from this book is for readers to begin to remember what it is to be human, to begin to remember what it is to be a member of a larger biotic community. What I want is for you—and me, and all of us—to fall back into the world into which you—and me, and all of us—were born, before you, too, like all of us were taught to become a bigot, before you, too, like all of us were taught to become a human supremacist, before you, too, like all of us were turned into a servant of this machine culture like your and my parents and their parents before them. I want for you—and me, and all of us—to fall into a world where you—like all of us—are one among many, a world of speaking subjects, a world of infinite complexity, a world where we each depend on the others, all of us understanding that the health of the real world is primary.

The world is being murdered. It is being murdered by actions that are perpetrated to support and perpetuate a worldview. Those actions must be stopped. Given what is at stake, failure is no longer an option. The truth is that it never was an option.

So where do we begin? We begin by questioning the unquestioned beliefs that are the real authorities of this culture, and then we move out from there. And once you've begun that questioning, my job is done, because once those questions start they never stop. From that point on, what you do is up to you.

EPILOGUE

An acquaintance read this manuscript, then said, "So much of what you say here makes sense, but I still have a big problem with it."

"What's that?" I responded.

"If agriculture and other authoritarian technics lead to overshoot and drawdown, which then lead to a choice of either collapse or conquest (and we know which this culture chooses); and if converting a land base to weapons of war leads to a short-term competitive advantage over cooperative and sustainable cultures; how do we stop them?"

I said, "Congratulations, and thank you."

He looked at me blankly.

I said, "You understood the point of the book far better than I could have dreamed."

He thought a moment, then asked the next question: "And what now?"

I responded, "Welcome to the war."

He looked at me for the longest time, then gave the barest—*barest*—hint of a smile. He nodded and said, "I'll see you on the front lines."

NOTES

1. I first encountered this as articulated by Robert Combs, *Vision of the Voyage*, (Memphis: Memphis State University Press, 1978), 2.

2. I'm with Mary Daly, who said that at this point there is (at least among the civilized) only one religion on the planet, and that is patriarchy. I am also with her (and Erich Fromm, Lewis Mumford, and others) when she talks about this culture's real religion as being necrophilia, or the love of death.

3. Lee Dye, "Why Mice Sing: Sex, and to Protect Habitat," *Yahoo News*, http://news.yahoo.com/why-mice-sing-sex-protect-habitat-105328182--abc-news-tech.html (accessed October 8, 2013).

4. Larry Pynn, "Acidification of Oceans Threatens to Change Entire Marine Ecosystem," *Vancouver Sun*, October 25, 2013, http://www.vancouversun.com/technology/Acidification+ oceans+threatens+change+entire+marine+ecosystem/9085021/story.html.

5. Rob Edwards, "Revealed: How Global Warming is Changing Scotland's Marine Life," *The Herald*, December 29, 2013, http://www.heraldscotland.com/news/home-news/revealed-how-global-warming-is-changing-scotlands-marine-life.23052108. (accessed December 31, 2013).

6. Joanna M. Foster, "Climate Change Will Starve the Deep Sea, Study Finds," *Think Progress*, January 2, 2014, http://thinkprogress.org/climate/2014/01/02/3113101/climate-change-starve-deep-sea/ (accessed January 3, 2014).

    Don't you love how commercial interests and scientific "discovery" are primary, and maintenance of life in the ocean rates an "also"?

7. "Salt-Water Fish Extinction Seen By 2048," *CBS News*, http://www.cbsnews.com/news/salt-water-fish-extinction-seen-by-2048/?mc_cid=f15c2bdadc&mc_eid=83a5da071d (accessed May 10, 2014).

8. Victoria Woollaston, "Mussels Could Soon Be Off The Menu: Climate Change May Wipe Out the Shellfish as Acid in Oceans Stops Their Shells Forming," *Daily Mail*, December 23, 2014, http://www.dailymail.co.uk/sciencetech/article-2885290/Mussels-soon-menu-climate-change.html (accessed December 26, 2014).

9. Michael Robotham, *Say You're Sorry* (New York: Little, Brown, 2012), 49-50. The book is a novel, but the description of sociopaths is accurate. The ellipses were in the original.

10. Except that, as we'll see, we haven't really gotten rid of God and the angels, but replaced them with Pure Reason and machines.

11. My thanks to Shelly Magnum for the previous paragraphs.

12. Originally at: "Probing Questions: Are Pigs Smarter Than Dogs?" at *Research Penn State*, but the page is gone. Here it is on archives: http://web.archive.org/

web/20060630042251/http://www.rps.psu.edu/probing/pigs.html (accessed November 25, 2013).

13. "Brain Evolution," *Your Amazing Brain*, http://www.youramazingbrain.org/insidebrain/ brainevolution.htm (accessed March 17, 2012). This stupid website isn't alone in its narcissism. This attitude is common. I just heard celebrity physicist Michio Kaku say the same thing in an interview with Jon Stewart.

14. The question and answer given in the article I'm citing are: "How Could Bees of Little Brain Come Up With Anything As Complex As a Dance Language? The answer could lie not in biology but in six-dimensional math and the bizarre world of quantum mechanics."

15. Declan Hayes, *God's Solution: Why Religion Not Science Answers Life's Deepest Questions* (Lincoln, Nebraska: Iuniverse, 2007), 255.

16. It ends up, and we'll discuss this a bit later, that between the time that Stamets said all of this and my writing this book, scientists have decided to no longer classify slime molds as fungi, but rather as amoebas who are single-celled when food is abundant, and who gather in communities and act as a single larger organism when food is more scarce. None of which alters how amazing they are.

17. A great example of this mimicry occurred to me today. I have a pretty bad flu infection, with coughing, fever, chills, the whole thing. I slept all day, waking up several times only long enough to stagger into the bathroom, use the toilet, and take more cough syrup. I don't know about you, but sometimes when I have these sorts of illnesses, I enter strange half-waking obsessions, where a single thought overtakes me and I won't be able to get rid of it. Today it was the song "Allouette," which I, like many others, learned as a child. I always presumed the song, which is in French, was about the adventures of a child named Allouette. But that's not true. It ends up that the song is about plucking a lark. It begins "Lark, nice lark/Lark, I shall pluck you/I shall pluck your head." And so on. The point is that I still remember this song from when I learned it forty-five years ago, but I had no idea what it was about. I learned it by mimicry. Does this mean I'm not intelligent?

18. The following TED talk is very interesting: Stefano Mancuso, "The Roots of Plant Intelligence," TED, 13:50, July 2010, http://www.ted.com/talks/stefano_mancuso_the_ roots_of_plant_intelligence.

19. "LINV at First Glance," International Laboratory of Plant Neurobiology, http://www. linv.org/linv_about.php (accessed October 27, 2013).

20. "Scientific Research Has Shown Plants Can Hear Themselves Being Eaten," news. com.au, July 3, 2014, http://www.news.com.au/technology/science/scientific-research- has-shown-plants-can-hear-themselves-being-eaten/story-fnjwkt0b-1226976987480 (accessed July 4, 2014).

21. Charles Darwin and Francis Darwin, *The Power of Movement in Plants* (London: John Murray, 1880), 572, http://www.gutenberg.org/cache/epub/5605/pg5605.html (accessed October 27, 2013)

22. Natalie Angier, "Sorry, Vegans: Brussels Sprouts Like to Live, Too," *The New York Times*, December 21, 2009, http://www.nytimes.com/2009/12/22/science/22angi.html?_r=0 (accessed October 27, 2013).

23. The proper word is *whom*.

24. I don't find it curious at all, but rather exactly what we should expect.

25. Please excuse the scare quotes. They were in the original. No human supremacist can conceive of plants having memories: instead, they only have "memories."

26. Which could, of course, be a mechanistic way to describe our memories as well. Of course I don't believe that our memories are *only* stored at the cellular level, but my point with this footnote is that if one does believe that this is what memories are, the same would be true for both animals (human and otherwise) and plants, which makes the scare quotes all the more out of place.

27. Michael Marder, "If Peas Can Talk, Should We Eat Them?" Opinionator, *The New York Times Online*, April 28, 2012, http://opinionator.blogs.nytimes.com/2012/04/28/if-peas-can-talk-should-we-eat-them/ (accessed May 8, 2012).

28. In my defense, I have Crohn's disease, which means I absorb food very poorly, so most of this wasn't being metabolized, but rather simply pooped out about three hours later.

29. Anthony Trewavas, "Green Plants as Intelligent Organisms," *Trends in Plant Science*, 10, no. 9 (September 2005): 413-419, http://www.linv.org/images/about_pdf/Trends%20 2005%20Trewavas.pdf (accessed October 27, 2013).

30. Ibid.

31. John H. Bodley and Foley C. Benson, "Stilt Root Walking by an Iriarteoid Palm in the Peruvian Amazon," *Biotropica*, 12, no. 1 (1980: 67, JSTOR Database, http://www.jstor. org/discover/10.2307/2387775?uid=3739560&uid=2129&uid=2&uid=70&uid=4&ui d=3739256&sid=21102823699901 (accessed October 27, 2013).

32. Trewavas, "Green Plants as Intelligent Organisms," 413–419.

33. Interestingly enough, according to Wikipedia, when fungi put haustoria into plants, the "host plant appears to be functioning according to signals from the fungus and the complex appears to be under the control of the invader." From "Haustorium," *Wikipedia*, http://www.wikipedi.org/wiki/Haustorium (accessed February 6, 2016).

34. Richard Alleyne, "Tomatoes Can 'Eat' Insects," *The Telegraph*, December 4, 2009, http://www.telegraph.co.uk/earth/wildlife/6727709/Tomatoes-can-eat-insects.html (accessed November 25, 2013).

35. Michael Pollan, "The Intelligent Plant: Scientists Debate a New Way of Understanding Flora," *The New Yorker*, December 23, 2013, http://www.newyorker.com/reporting/2013/12/23/131223fa_fact_pollan?currentPage=all (accessed December 31, 2013).

36. The *Wired* article is Mark Brown, "Fish Photographed Using Tools to Eat," *Wired*, July 11, 2011, http://www.wired.com/wiredscience/2011/07/fish-tool-use/ (accessed January 6, 2014).

    The *Field & Stream* article is Chad Love, "Fish Learn to Use Tools," *Field & Stream*, July 11, 2011, http://www.fieldandstream.com/blogs/field-notes/2011/07/rise-planet-fish-theyve-learned-use-tools (accessed January 6, 2014).

37. Joh R. Henschel, "Tool Use by Spiders: Stone Selection and Placement by Corolla Spiders *Ariadna* (Segestriidae) of the Namib Desert," *Ethology*, 1, no. 3 (January–December 1995): 187–199, http://onlinelibrary.wiley.com/doi/10.1111/j.1439-0310.1995. tb00357.x/abstract (accessed January 6, 2014).

38. Ker Than, "Wooly Bear Caterpillars Self-Medicate: A Bug First," *National Geographic News*, March 13, 2009, http://news.nationalgeographic.com/news/2009/03/090313-self-medicating-caterpillars.html?source=rss (accessed September 4, 2014).

39.   http://blogs.warwick.ac.uk/avedgeworth/?num=10&start=10 (accessed January 4, 2014).

40.   Neil Evernden, *The Natural Alien: Humankind and Environment* (Toronto: University of Toronto Press, 1985), 16–17.

41.   This is all so similar to what the person said about slugs. If trees share one of our traits, not good for the ego, eh? Well, not if your ego is so fragile as to require you to be "more special" than everyone else.

42.   Kat McGowan, "How Plants Secretly Talk to Each Other," *Wired*, December 20, 2013, http://www.wired.com/wiredscience/2013/12/secret-language-of-plants/ (accessed January 20, 2014).

43.   Ibid. And can you imagine how horribly various pollutants must affect their ability to perceive?

44.   Ibid.

45.   Pollan, "The Intelligent Plant."

46.   "Teleology," *Oxford Dictionaries*, http://www.oxforddictionaries.com/us/definition/american_english/teleology (accessed January 21, 2014).

47.   "Define Teleology," *AskDefine*, http://teleology.askdefine.com/ (accessed January 21, 2014).

48.   Bob Drury, "Darwinism Can Survive Without Teleology," *Catholic Stand*, http://catholicstand.com/evolution-design/ (accessed January 21, 2014).

49.   For an enlightening exploration of Francis Bacon and his hatred of both nature and women, see Carolyn Merchant, *The Death of Nature: Women, Ecology, and the Scientific Revolution* San Francisco: HarperSanFrancisco, 1990), especially pages 64–91.

50.   Marianela Jarroud, "Where Would You Like Your New Glacier?" Inter Press Service, http://www.ipsnews.net/2014/02/like-new-glacier/ (accessed March 4, 2014).

51.   Pollan, "The Intelligent Plant."

52.   Ibid.

53.   Ibid.

54.   McGowan, "How Plants Secretly Talk to Each Other."

55.   Mann has also spoken in favor of the oil and gas industry.

56.   Standing dead trees are often homes for more creatures than are live trees.

57.   "Survival of the fittest" is actually Herbert Spencer's term, but Darwin later adopted it.

58.   *Mother Tree Sanctuary*, http://www.mothertreesanctuary.org/#!mother-trees/c1l40 (accessed March 3, 2014).

59.   Jane Engelsiepen, "Trees Communicate With Each Other," *Positive News*, http://www.positivenewsus.org/trees-communicate-with-each-other.html (accessed March 3, 2014).

60.   Does anyone else find his choice of names significant in terms of his worldview?

61.   Richard Dawkins, *The Selfish Gene* (Oxford: Oxford University Press, 1989), 184–185.

62.   Geoffrey Mohan, "Chimpanzees Make Monkeys of Humans in Computer Game," *Los Angeles Times*, June 6, 2014, http://www.latimes.com/science/sciencenow/la-sci-sn-chimp-game-theory-humans-20140605-story.html (accessed June 11, 2014).

63.   McGowan, "How Plants Secretly Talk to Each Other."

64.   To eat is not the same as to attack.

65.   Pollan, "The Intelligent Plant."

66. Please recall the previous footnote about Paul Stamets calling slime molds fungi; between when Stamets said that and when Pollan wrote the current article, scientists changed their classification of slime molds.

67. Pollan, "The Intelligent Plant."

68. Heather Smith, "Want Everyone Else to Buy Into Environmentalism? Never Say 'Earth,'" *Grist*, http://grist.org/climate-energy/want-everyone-else-to-buy-into-environmentalism-never-say-earth/ (accessed March 17, 2014).

    Also, Peter Kareiva, R. Lalasz, and M. Marvier, "Conservation in the Anthropocene: Beyond Solitude and Fragility," *Breakthrough Journal*, 1, no. 3 (Winter 2012): 36.

69. William J. Broad, "Billionaires With Big Ideas Are Privatizing Science," *The New York Times*, March 15, 2104, http://www.nytimes.com/2014/03/16/science/billionaires-with-big-ideas-are-privatizing-american-science.html?_r=0 (accessed March 17, 2014).

70. Leonard Finkleman, "Peas and Quiet," *Rationally Speaking*, March 11, 2012, http://rationallyspeaking.blogspot.com/2012/05/peas-and-quiet.html (accessed March 19, 2014).

71. For example, the comedian Alan Davies lost ten points for saying that Julius Caesar was born by Cesarian section. I'd always heard that, but it ends up it's not true. At that time, the women always died in the operation, but Caesar's mother lived into his twenties, meaning he couldn't possibly have been born this way. I know we've also been told this is the etymology of the word *Cesarian*, but see Wikipedia for an interesting discussion: "Caesarian Section," http://en.wikipedia.org/wiki/Caesarean_section (accessed March 23, 2014).

72. *QI*, Series B, episode 6, "Beavers," http://www.youtube.com/watch?v=4Njfk0qqM24 (accessed March 21, 2014).

73. "Bacteria Communicate to Help Each Other Resist Antibiotics," *Science Daily*, July 4, 2013, http://www.sciencedaily.com/releases/2013/07/130704095130.htm (accessed March 25, 2014).

    I find it extremely interesting that scientists seem to have less difficulty using words like "communicate," "language," and "help" concerning bacteria than they do concerning plants.

74. Melinda Wenner, "Quiet Bacteria and Antibiotic Resistance," *Scientific American*, April 20, 2009, http://www.scientificamerican.com/article/bacteria-antibiotic-resistance/ (accessed March 25, 2014).

75. With about 1,400 different strains of bacteria living just in your navel.

76. Vittorio Venturi and Sujatha Subramoni, "Future Research Trends in the Major Chemical Language of Bacteria," *HFSP Journal*, 3, no 2 (April 2009): 105–116, published online March 4, 2009, doi: 10.2976/1.3065673; United States Library of Medicine, National Institutes of Health, http://www.ncbi.nlm.nih.gov/pmc/articles/PMC2707791/ (accessed March 26, 2014).

77. Anthony Trewavas, "The Green Plant as an Intelligent Organism," in *Communication in Plants: Neuronal Aspects of Plant Life*, ed. Frantisek Baluska, Stefano Mancuso, and Dieter Volkmann (New York: Springer, 2006), 6.

78. Cobbled together from a number of websites, including, among others, "Tardigrade," *Wikipedia*, http://en.wikipedia.org/wiki/Tardigrade;

Helen Pow, "Meet the Toughest Animal on the Planet," *Daily Mail*, http://www.dailymail.co.uk/news/article-2280286/Meet-toughest-animal-planet-The-water-bear-survive-frozen-boiled-float-space-live-200-years.html

"How Long Does a Tardigrade Live?" Fluther.com, http://www.fluther.com/3530/how-long-does-a-tardigrade-live/ (accessed May 10, 2014).

79. Traci Watson, "Bdelloids Surviving on Borrowed DNA, *Science Now*, November 15, 2012, http://news.sciencemag.org/evolution/2012/11/bdelloids-surviving-borrowed-dna. Site visited (accessed May 10, 2014).

80. John Graham Dalyell, *Observations on Some Interesting Phenomena in Animal Physiology, Exhibited by Several Species of Planariae: Illustrated by Coloured Figures of Living Animals* (Edinburgh: The Archibald Constable & Co., 1814), 32, http://www2u.biglobe.ne.jp/~gen-yu/pla_classic_e.html (accessed May 1, 2014).

81. Teresa Coppens, "What is Negligible Senescence?" http://hubpages.com/hub/What-is-Negligible-Senescence (accessed May 14, 2014).

82. Ker Than, "'Immortal' Jellyfish Swarms World Oceans," *National Geographic*, January 29, 2009, http://news.nationalgeographic.com/news/2009/01/090130-immortal-jelly-fish-swarm.html (accessed May 14, 2014).

83. Ibid.

84. And how do they know this? Because they injected acid into the mole rats' paws.

85. As well as many other religions. Some say it is the origin of all religions.

86. Unless, of course, the religion considers life in heaven to be, well, heaven; and life on earth to be hell, in which case planarians, lobsters, and bacteria are clearly sinners being punished; but I can't imagine how any religion would be ridiculous enough to consider this wonderful gift that is life on earth to be even remotely hellish.

87. Or there's this not atypically narcissistic language from "futurist" Jason Silva, who says, "The mindset of an Immortalist is simple and straightforward: death is an abhorrent imposition on a species able to reflect and care about meaning. Creatures that love and dream and create and yearn for something meaningful, eternal and transcendent should not have to suffer despair, decay and death. We are the arbiters of value in an otherwise meaningless universe. The fleeting nature of beautiful, transcendent moments feeds the urge for man to scream: 'I was here; I felt this and it matters, goddamn it!' In the face of meaningless extinction, it's not surprising that mankind has needed to find a justification for his suffering. Man is the only animal aware of his mortality—and this awareness causes a tremendous amount of anxiety."

In the same essay, he writes, "The time has come for man to get over his cosmic inferiority complex. To rise above his condition—to use technology to extend himself beyond his biological limitations. Alan Harrington reminds us: 'We must never forget we are cosmic revolutionaries, not stooges conscripted to advance a natural order that kills everybody.' While Ernest Becker identified our need for heroism and our extensive attempts to satisfy it symbolically, Alan Harrington proposes we move definitively to engineer salvation in the real world; to move directly to physically overcome death itself: 'Spend the money, hire the scientists and hunt down death like an outlaw.'

He concludes, "The Immortalist point of view, then, could be described as a project that uses technology to 'Individualize eternity, to stabilize the forms and identities through which the energy of conscious life passes.' This is hardly a stretch for human beings, as Harrington proclaims: 'We have long since gone beyond the moon, touched down on mars, harnessed nuclear energy, artificially reproduced DNA, and now have the biochemical means to control birth; why should death itself, "the last enemy," be considered beyond conquest?'"

From Ian Mackenzie, "The End of Death: Further Conversations with Jason Silva," *Matadornetwork*, August 6, 2009, http://matadornetwork.com/bnt/the-end-of-death-further-conversations-with-jason-silva/ (accessed January 23, 2015).

88. Alex Kirby, "How Whale Feces Slows Down Global Warming," *Alternet*, http://www.alternet.org/environment/whale-feces-helps-slow-antarctic-warming?akid=11658.202899.PjfwM4&rd=1&src=newsletter976470&t=25&paging=off&current_page=1#bookmark (accessed April 6, 2014).

89. "Ernest Duchesne," *Wikipedia*, http://en.wikipedia.org/wiki/Ernest_Duchesne (accessed June 3, 2014).

90. I of course would never do any of these, but you get the picture.

91. "*Fantasia*," *IMDB*, http://www.imdb.com/title/tt0032455/trivia (accessed June 15, 2014).

92. David Fazekas, "What If Mosquitoes Were Annihilated?" *Yahoo*, May 29, 2014, http://news.yahoo.com/blogs/what-if-abc-news/what-if-mosquitoes-were-anihilated-194753142.html (accessed June 3, 2014).

Two more things. The first is that nearly all of the comments below the article were in favor of this, simply because a lot of people find mosquitoes annoying. Only a few of the commenters were sane, describing what could happen to bats or others who consume mosquitoes, and describing the roles that mosquitoes play in natural communities. The other is that when I went to cut and paste the first quote from the article, there was a macro in the text that destroyed my entire book manuscript. Fortunately, and only because I've lost so much text so many times over the years, I belong to the religion of "backing up my files." My point in bringing that up is that if something as simple as cutting and pasting text can destroy a manuscript, what can happen when you decisively interfere in something much larger and more complex, *and living*, like a forest? And further, if I understand the necessity of backing up manuscripts before making major changes, why can't more people understand the necessity of not making large irreversible changes to the real world?

93. "Oxytech 'Negligent' Over GM Mosquito Release in Panama," *Sustainable Pulse*, February 12, 2014, http://sustainablepulse.com/2014/02/12/oxitec-negligent-gm-mosquito-release-panama/#.U44i1z9OXmQ (accessed June 3, 2014).

94. Clare Wilson, "Baby's First Gut Bacteria Could Come from Mum's Mouth," *New Scientist*, May 21, 2014, http://www.newscientist.com/article/dn25603-babys-first-gut-bacteria-may-come-from-mums-mouth.html#.U8rY3L7n_VI (accessed July 19, 2014).

95. Jonathan Colob, "What is Snot?" *The Stranger*, December 21, 2011, http://www.thestranger.com/seattle/dear-science/Content?oid=11182785 (accessed June 4, 2014).

96. Mary Bates, "Rats Regret Making the Wrong Decision," *Wired*, June 8, 2014, http://www.wired.com/2014/06/rats-regret-making-the-wrong-decision/ (accessed June 18, 2014).

97. Ann Koh and Heesu Lee, "Arctic Ice Melt Seen Freeing Way for South Korean Oil Hub," *Bloomberg*, July 23, 2014, http://www.bloomberg.com/news/2014-07-22/arctic-ice-melt-seen-freeing-way-for-south-korean-oil-hub.html (accessed July 23, 2014).

98. Terrell Johnson, "Climate Change Tourism Comes to the Arctic: $20,000 Luxury Cruise to Sail the Once-Unnavigable Northwest Passage," weather.com, July 29, 2014, http://www.weather.com/news/science/environment/melting-arctic-20000-luxury-cruises-climate-change-tourism-20140728 (accessed July 31, 2014).

99. *Johnson v M'Intosh* (1823), http://scholar.google.com/scholar_case?-case=3104237999990733260&q=johnson+v.+mcintosh&hl=en&as_sdt=2006&as_vis=1 (accessed June 19, 2014).

100. And don't give me any shit about how you could use your credit card to buy ladybugs, which will be delivered to your door and then dumped on your head; this means you have to rely on money, which means it's only available to the financial elite. This is typical of many of the "miracles" of modern society: they increase our dependence on the economic and political system, as opposed to the real world.

101. Dawkins, *The Selfish Gene*.

102. Kendall F. Haven, *100 Greatest Science Inventions of All Time* (Santa Barbara: Libraries Unlimited, 2005), 18.

103. Nic Fleming, "'Cunning' Squirrels Pretend to Bury Their Food," *The Telegraph*, January 16, 2008, http://www.telegraph.co.uk/science/science-news/3322101/Cunning-squirrels-pretend-to-bury-their-food.html (accessed July 6, 2014).

    Many other species also store food in ways that help the landbase. For example, "When food is plentiful, late summer and fall, the chickadees hoard food. They stash food under bark or in patches of lichen. A single chickadee may stockpile hundreds of food items in a day; placing each item in a different spot. Chickadees can remember thousands of food hiding places. They can retrieve the food item with almost perfect accuracy 24 hours later. Some birds can even remember the location of their cache for up to 28 days after hiding." From "Black-Capped Chickadee," *Holdenarb*, http://www.holdenarb.org/resources/Black-cappedChickadee.asp (accessed December 10, 2014).

104. James Fallows, "The 50 Greatest Breakthroughs Since the Wheel," *The Atlantic*, October 23, 2013, http://www.theatlantic.com/magazine/archive/2013/11/innovations-list/309536/ (accessed July 4, 2014).

105. And at this point in the book I hope you don't suggest that just because humans can't perceive some sense, that it does not exist.

106. Bill Frezza, "In The Battle of Man Vs. Nature, Give Me Man," *Forbes*, January 3, 2012, http://www.forbes.com/sites/billfrezza/2012/01/03/in-the-battle-of-man-vs-nature-give-me-man/ (accessed June 25, 2014).

107. Julie Cart, "Sacrificing the Desert to Save the Earth," *Los Angeles Times*, February 5, 2012, http://www.latimes.com/news/local/la-me-solar-desert-20120205,0,7889582.story (accessed July 2, 2014).

108. Others, like Guy McPherson, also recognize the destructiveness of this culture, and that it will likely result in human extinction. I don't mention him in the text because he is not a human supremacist, and is fighting desperately to help collapse industrial civilization.

109. Eli Strokols, "Judge Overturns Fort Collins Five-year Fracking Ban," *Fox31 Denver*, August 7, 2014, http://kdvr.com/2014/08/07/judge-overturns-fort-collins-five-year-fracking-ban/?mc_cid=b6a58e25f9&mc_eid=83a5da071d (accessed August 9, 2014).

110. Lewis Mumford, "Authoritarian and Democratic Technics," *Technology and Culture*, 5, no. 1 (Winter 1964): 2.

111. Mumford, "Authoritarian and Democratic Technics," 3–4.

112. G. K. Chesterton, *The Complete Father Brown Stories* (London: BBC Books, 2013), 84.

113. Once again, how do you think the world became so fecund and beautiful in the first place, if not by the actions of those who live here? Contrary to the beliefs of human supremacists of flavors both monotheistic and scientific, the world didn't just *somehow* become this wonderful all so we can destroy it. As accustomed as I unfortunately am to this culture's insanity, it still never ceases to amaze me that while almost no sane people of good heart believe the capitalist conceit that selfish individuals ruthlessly trying to exploit each other will, through the magic of Adam Smith's Invisible Hand, lead to healthy, functioning, vibrant communities, yet a lot of seemingly sane scientists of seemingly good heart can without flinching project the same nonsensical capitalist conceit onto the natural world and believe that selfish individuals ruthlessly trying to exploit each other will, this time through the magic of random actions, lead to healthy functioning natural communities, or as they call them, "ecosystems."

114. Mumford, "Authoritarian and Democratic Technics," 4.

115. Mumford, "Authoritarian and Democratic Technics," 4–5.

116. Noam Chomsky, "Human Intelligence and the Environment," *International Socialist Review* 76, based on a talk he gave September 30, 2010. http://isreview.org/issue/76/human-intelligence-and-environment (accessed August 27, 2014).

117. Jason Patinkin, "Kenya Conundrum: Kick Out Maasai Herders to Develop Geothermal Energy?" *Christian Science Monitor*, September 10, 2014, http://www.csmonitor.com/World/Africa/2014/0910/Kenya-conundrum-Kick-out-Masai-herders-to-develop-geo-thermal-energy (accessed September 27, 2014).

118. Toby Hemenway, "Is Sustainable Agriculture an Oxymoron?" *Pattern Literacy*, http://www.patternliteracy.com/203-is-sustainable-agriculture-an-oxymoron (accessed September 25, 2014).
    I really hope you read the whole essay. It's fantastic.

119. Maria Mies, *Patriarchy and Accumulation on a World Scale* (London: Zed Books, 1999), 98.

120. I have experienced this directly. I hit an overturned flatbed load of plywood going fifty-five miles per hour. I walked away. My mother broke her neck. Had I been going sixty-five, would she have died? Had I been going forty-five, would she have walked away?

121. Mumford, "Authoritarian and Democratic Technics," 5.

122. Ibid.

123. Mumford, "Authoritarian and Democratic Technics," 5–6.

124. Mumford, "Authoritarian and Democratic Technics," 6.

125. Lewis Mumford, Address to the National Book Awards Committee, in Donald L. Miller, *Lewis Mumford: A Life* (New York, Grove Press, 2002).

126. Mumford, "Authoritarian and Democratic Technics," 7.

127. Michael Wines, "In Alaska, a Battle to Keep Trees, or an Industry, Standing," *The New York Times*, September 27, 2014, http://www.nytimes.com/2014/09/28/us/a-battle-to-keep-trees-or-an-industry-standing.html (accessed October 5, 2014).

128. Alanna Mitchell, "Winged Warnings: Built for Survival, Birds in Trouble From Pole to Pole," *Truthout*, September 30, 2014, http://truth-out.org/news/item/26514-winged-

warnings-built-for-survival-birds-in-trouble-from-pole-to-pole (accessed October 5, 2014).

129. Tracy McVeigh, "In the Age of Extinction, Which Species Can We Least Afford to Lose?" *The Guardian*, October 4, 2014, http://www.theguardian.com/environment/2014/oct/05/threatened-species-cannot-afford-to-lose-age-of-extinction (accessed October 5, 2014).

130. It's interesting, the word processing program *Word* contains in its dictionary the term *Anthropocene* and at the same time marks as grammatically incorrect the use of the word *who* when applied to nonhumans. The human supremacism is everywhere.

131. Jim Robbins, "Building an Ark for the Anthropocene," *New York Times*, September 27, 2014, http://www.nytimes.com/2014/09/28/sunday-review/building-an-ark-for-the-anthropocene.html (accessed October 11, 2014).

132. These are all real articles in *The New York Times*.

133. Edward Gibbon, *The Decline and Fall of the Roman Empire*, Volume 3, The Modern Library (New York: Random House, no date given), 665.

134. Ibid., 657.

135. Elizabeth Douglass, "Q&A With SolarCity's Chief: There Is No Cost to Solar Energy, Only Savings," *Inside Climate News*, October 17, 2014, http://insideclimatenews.org/news/20141017/qa-solarcitys-chief-there-no-cost-solar-energy-only-savings (accessed November 4, 2014).

136. This paragraph is mostly put together from:
"Rare-Earth Mining in China Comes at a Heavy Cost for Local Villages: Pollution is Poisoning the Farms and Villages of the Region That Processes the Precious Minerals," *The Guardian*, August 7, 2012, http://www.theguardian.com/environment/2012/aug/07/china-rare-earth-village-pollution (accessed November 4, 2014).
Jonathan Kaiman, "Rare Earth Mining in China: The Bleak Social and Environmental Costs," *The Guardian*, March 20, 2014, http://www.theguardian.com/sustainable-business/rare-earth-mining-china-social-environmental-costs (accessed November 4, 2014).

137. Max Wilbert, personal communication, October 30, 2014.

138. From Ozzie Zehner, *Green Illusions: The Dirty Secrets of Clean Energy and the Future of Environmentalism* (Lincoln: University of Nebraska Press, 2012).

139. "Toward a Just and Sustainable Solar Energy Industry," *Silicon Valley Toxics Coalition*, January 2009, page 1, http://svtc.org/wp-content/uploads/Silicon_Valley_Toxics_Coalition_-_Toward_a_Just_and_Sust1.pdf (accessed November 7, 2014).

140. And keep in mind the contempt and hatred that necessarily accompanies any supremacism, and the contempt and hatred of nature that necessarily accompanies human supremacism.

141. Ross Anderson, "Exodus: Elon Musk Argues that We Must Put a Million People On Mars if We Are to Ensure That Humanity Has a Future," *Aeon*, http://aeon.co/magazine/technology/the-elon-musk-interview-on-mars/ (accessed November 13, 2014).

142. Mark Koba, "Trees vs. Humans: In California Drought, Nature Gets to Water First," CNBC, October 17, 2014, http://www.cnbc.com/id/102094056# (accessed December 1, 2014).
At least the public comments recognize that the piece is timber industry propaganda.

143. Ari Phillips, "Texas' Top Toxicologist: EPA's New Smog Regulations Unnecessary, Just Stay Indoors," *Climate Progress*, October 21, 2014, http://thinkprogress.org/climate/2014/10/21/3582548/smog-people-stay-indoors-anyway/ (accessed December 2, 2014).

144. Mike Ives, "Drive to Mine the Deep Sea Raises Concerns Over Impacts," October 20, 2014, http://e360.yale.edu/feature/drive_to_mine_the_deep_sea_raises_concerns_over_impacts/2818/ (accessed December 2, 2014).

145. Josh Schlossberg, "Public Forests Sacrificed to the Biomass Industry," *EcoWatch: Transforming Green*, October 20, 2014, http://ecowatch.com/2014/10/20/national-forests-sacrificed-biomass-industry/ (accessed December 2, 2014).

146. Robin McKie, "After 42 Years of Charting the Health of Our Seas, Scientist's Studies Now Face the Axe," *The Guardian,* October 25, 2014, http://www.theguardian.com/environment/2014/oct/26/guillemots-study-skomer-wales-budget-cut-tim-birkhead (accessed December 2, 2014).

147. Rachael Bale, "The Surprising Reason Abandoned US Mines Haven't Been Cleaned Up," *The Center For Investigative Reporting*, November 4, 2014, https://beta.cironline.org/reports/the-surprising-reason-abandoned-us-mines-havent-been-cleaned-up/ (accessed December 2, 2014).

148. Nancy Lofholm, "Dead Babies Near Oil Drilling Sites Raise Questions For Researchers," *The Denver Post*, October 26, 2014, http://www.denverpost.com/news/ci_26800380/dead-babies-near-oil-drilling-sites-raise-questions (accessed December 2, 2014).

149. Mark Koba, "Fracking or Drinking Water? That May Become the Choice," *NBC News*, September 14, 2014, http://www.nbcnews.com/business/business-news/fracking-or-drinking-water-may-become-choice-n202231 (accessed December 2, 2014).

150. Brad Plumer, "We're Damming Up Every Last Big River on Earth. Is That Really a Good Idea?" *Vox*, October 28, 2014, http://www.vox.com/2014/10/28/7083487/the-world-is-building-thousands-of-new-dams-is-that-really-a-good-idea (accessed December 2, 2014).

151. Bobby Magill, "Hydropower May Be Huge Source of Methane Emissions," *Climate Central*, October 29, 2014, http://www.climatecentral.org/news/hydropower-as-major-methane-emitter-18246 (accessed December 2, 2014).

152. Jonathan Watts, "Amazon Rainforest Losing Ability to Regulate Climate, Scientist Warns," *The Guardian*, October 31, 2014, http://www.theguardian.com/environment/2014/oct/31/amazon-rainforest-deforestation-weather-droughts-report (accessed December 2, 2014).

153. Jonathan Watts, "Amazon Deforestation Picking Up Pace, Satellite Data Reveals: Data Indicates," *The Guardian*, October 19, 2014, http://www.theguardian.com/environment/2014/oct/19/amazon-deforestation-satellite-data-brazil (accessed December 2, 2014).

154. "Giant Palouse Earthworm Press Release," University of Idaho, April 27 (unstated year), http://www.uidaho.edu/cals/news/feature/gpe/pressrelease (accessed October 19, 2014).
    The scientists now say the worms only grow to one foot, and don't smell like lilies. But the worms may grow larger and smell different in the real world than they do in a lab.

155. "Black Bear Eats Body of Californian Man," *The Sun Daily*, October 18, 2014, http://www.thesundaily.my/news/1202499 (accessed December 7, 2014).

And I just got a note from a friend of mine whose good friend was close to the man who died. She said, "He would have thought being eaten by a bear after he died was perfect and wonderful." We should all be so fortunate as to be able to give back like that after we die. That was the gift that everyone gave and received, until recently, and that's how it's supposed to be.

156. R. D. Laing, *The Politics of Experience* (New York: Ballantine Books, 1967), 78.

157. "Me-Me-Og, The Wild Pigeon of North America," *Hunter-Trader-Trapper* XVI, no. 3 (June 1908): 48, http://books.google.com/books?id=biTOAAAAMAAJ&pg=PA47&lpg=PA47&dq=%22I+have+seen+them+move+in+one+unbroken+column+-for+hours+across+the+sky,+like+some+great+river,+ever+varying+in+hue%22&-source=bl&ots=H03PuvDXnT&sig=lxCmuWJPQ_f8Llx8xHdHWKSxLx-Q&hl=en&sa=X&ei=6M2EVKP3BMvmoASR94HgCQ&ved=0CCAQ6A-EwAA#v=onepage&q=%22I%20have%20seen%20them%20move%20in%20one%20unbroken%20column%20for%20hours%20across%20the%20sky%2C%20like%20some%20great%20river%2C%20ever%20varying%20in%20hue%22&f=false (accessed December 7, 2014).

The entire account is breathtaking and heartbreaking. The author states, for example, that the flocks sounded like "the strange commingling sounds of sleigh bells, mixed with the rumbling of an approaching storm." The author's reverential perspective contrasts sharply with that of overt human supremacists like Charles Mann, whom we met earlier, who approvingly cites someone as saying the birds were "incredibly dumb." As I say in *Endgame*, the ones who were incredibly stupid were those who eradicated them.

158. Margaret Sarah Fuller, *Summer on the Lakes, in 1843* (Champaign: University of Illinois Press, 1990), 39, http://books.google.com/books?id=mxA7wvjpt5UC&pg=PA39&lpg=PA39&dq=%22came+sweeping+across+the+lawn,+positively+in+-clouds,+and+with+a+swiftness+and+softness+of+winged+motion,+more+beauti-ful+than+anything+of+the+kind+I+ever+knew.%22&source=bl&ots=Moz5Wjtk-gr&sig=4M_6mCeJ0mGkk2Nt1ug6ZZe9WGY&hl=en&sa=X&ei=QdGEVJz8OM-W0oQSdl4GYDg&ved=0CCYQ6AEwAg#v=onepage&q=%22came%20sweeping%20across%20the%20lawn%2C%20positively%20in%20clouds%2C%20and%20with%20a%20swiftness%20and%20softness%20of%20winged%20motion%2C%20more%20beautiful%20than%20anything%20of%20the%20kind%20I%20ever%20knew.%22&f=false (accessed December 7, 2014).

159. Charles J. Hanley, "As Climate Talks Drag On, United Nations Scientists Ponder Use of Climate Engineering," AP, December 5, 2010.

160. I like this last phrase.

161. Erin Biba, "Science to the Rescue: Rebooting the Planet," *Newsweek*, December 4, 2014, http://www.newsweek.com/2014/12/12/can-geoengineering-save-earth-289124.html (accessed December 15, 2014).

162. Joe Romm, "Geoengineering Gone Wild: Newsweek Touts Turning Humans Into Hob-bits To Save Climate," *Climate Progress*, December 5, 2014, http://thinkprogress.org/climate/2014/12/05/3599762/geoengineering-newsweek-hobbits/?mc_cid=a078a37b-9b&mc_eid=83a5da071d (accessed December 15, 2014).

163. That energy efficiency leads not to energy savings, but increased consumption, was first described in the 1860s by British economist William Stanley Jevons. It is now called the Jevons Paradox.

164. And of course, coming up with this theory or model in the first place involved, ahem, speculating. Oh, the unscientific horror!

165. Between the time I wrote this book and the final edits before going to press, the number declined from five to three.

princet⌂n